3D Cell Culture

3D Cell Culture

**Fundamentals and Applications in Tissue
Engineering and Regenerative Medicine**

Ranjna C. Dutta | Aroop K. Dutta

PAN STANFORD PUBLISHING

Published by

Pan Stanford Publishing Pte. Ltd.
Penthouse Level, Suntec Tower 3
8 Temasek Boulevard
Singapore 038988

Email: editorial@panstanford.com
Web: www.panstanford.com

British Library Cataloguing-in-Publication Data
A catalogue record for this book is available from the British Library.

3D Cell Culture: Fundamentals and Applications in Tissue Engineering and Regenerative Medicine

ISBN 978-981-4774-53-6 (Hardcover)
ISBN 978-1-315-14682-9 (eBook)

Contents

Preface

Cell culture is an in vitro laboratory technique widely used for growing plant and animal cells. It is primarily utilized to learn and evaluate cellular processes under artificial but defined experimental conditions. Isolated cells and cell lines are enabled to grow ex vivo in glass or plastic appliances using artificial nutrients and supplements for targeted explorations. The composition of liquid medium is adjusted to meet the requirement of different types of cells to be cultured. Till now culturing cells in a flat-bottomed flask or Petri dish where cells are allowed to expand on a flat surface, i.e., in two dimensions has been a common practice. The technique for its ease is now widely adopted in clinics for diagnostic purposes. Use of cell culture techniques has not remained limited to unraveling the mysteries of cell biology but also extended to other areas like qualitative and quantitative assessment of metabolic activities.

The possibility of growing healthy and viable cells ex vivo led to the human urge towards a new direction for achieving longevity by replacing the ailing or diseased body tissues with renewed-healthy cells and tissues. Organ transplantation has been viewed as the most preferred choice in the state of kidney and heart failures. Tumorous tissues also leave no choice but to be removed and if possible replace the organ with the normal healthier ones. Replacement with healthy normal tissues could be a better choice for cancerous tissue before metastasis. Even after the onset of metastasis a swapping with healthy tissue might provide a control as the source of malignant signaling would not be there. Though, identifying an accurate match and a willing donor has always been scarce. This led to an expedition for generating ex vivo implants by growing patients' own cells. Such artificially created functional tissue is expected to replace the diseased tissue/organ without any fear of rejection. The quest was stumbled upon by the fact that artificially grown cells by conventional means are functionally unviable. Mostly they grow and remain as monolayer, hardly rearranging and adapting a cohesive tissue like features. It is realized that the conventional culture methods produce the cells that may morphologically look alike their counterpart present in vivo but almost always are functionally compromised. In order to produce physiologically viable cells, replicating in vivo like physical environment by providing three dimensional (3D) spaces for the cells to grow is attempted. This could help in improving the functional performance of the cells while still necessitating conventional supplements. Though, culturing cells in 3D space created through inert conventional polymers could produce relatively better results than those achieved by growing them in 2D culture wares yet they can barely yield physiologically viable tissue. It is speculated therefore that a 3D scaffold created using a biochemical mimic of in vivo like tissue specific extracellular matrix (ECM) would lead to far more gratifying outcomes. Thus, for generating functional cells applicable for tissue engineering and regenerative medicine purposes we need devices that mimic ECM both physically and biochemically. Not only that, we also have to improve our understanding about cell-ECM dynamics and interdependency of different cells. Cells need to be co-cultured for achieving desired co-operative and co-ordinated physiological outcome. Till now cells have generally been evaluated in isolation without considering the fact that they would behave differently if other cells are present in their surroundings, which is the case in native tissue environment.

We are at the onset of a leap by switching over from 2D to 3D cell-culture practices. Though there are limited devices available at present for culturing cells in the desired ECM mimicking environment provided in 3D, we still hope for better accessibility to such 3D culture wares in the near future. With the enormous efforts all over the world towards culturing cells in a tissue mimicking environment we keenly anticipate vital improvements in our understanding of cell-biology, diagnosis and tissue engineering. Better physiological control over artificially grown cells will certainly have its impact on the efficiency and acceptability of regenerative medicine.

Acknowledgements

A piece of work like this is a reflection of incremental acquisition of understanding which is not possible by an individual's lone efforts in isolation. Therefore I deem it my duty to acknowledge my Ph.D. guide Padam Shri Dr. Nitya Nand, mentors Drs. CM Gupta, NM Khanna (late), SK Basu, DM Salunke, GP Talwar, KV Raghavan, B Sesikeran, Dr. Sundaresh and Prof. Erwin Goldberg. Colleagues present and in the past that include Drs. VML Srivastava, V Bhakuni (late), SP Singh, A Puri, RA Vishwakarma, A Mukhopadhyay, K Balasubramanian, S Garg, K Kaur, S Shah, AK Panda, N Ehtesham, S Ghosh and P Baligar. Profs. B Basu and S Bose of IISc, Bangalore and NW Fadnavis of IICT, Hyderabad deserve special mention for their generous help and support. We are pleased to acknowledge each of the students, trainees and associates who contributed in our projects. Also appreciate the blessings and gracious co-operation of our immediate and extended family.

Last but not the least we wish to record our thanks to the funding agencies like DBT, DSIR, DST, India and acknowledge the efforts of all the researchers and scientists working relentlessly towards developing innovative solutions so that the pain and diseases could be minimised or eliminated in fellow humans either through new drugs or engineered or regenerated functional tissue. They may not directly be involved in putting this work together but their contribution to the field is equally worthy and important.

Synopsis

This work is an attempt to bridge the knowledge gaps between Medical need and Technology applications in a step-wise manner through illustrations. Available models for 3D cell culture as well as techniques to create substrates and scaffolds to achieve a desired 3D microenvironment are discussed. 3D cell culture has yet to be adopted and exploited to its full potential. It promises to upgrade and bring our understanding of human physiology to the highest level with the scope of applying the knowledge for better diagnosis as well as therapeutics.

The first chapter of the book deals with the technique of in vitro cell culture. A brief history and how the procedure has been adopted to its present form are described. The importance the procedure has gained over the years and its impact in drug development and diagnostics are also analyzed. Developments in the direction of 3D cell culture where another dimension is provided physically to growing cells and which ultimately revealed the limitations associated with flat bottom culture are explained.

The second chapter is dedicated to developing an understanding of the microenvironment that a cell or tissue is exposed to naturally, in a physiologically viable system. An attempt is made to establish the significance of the extracellular microenvironment (ECM) in terms of its architecture, mechanics and biochemical behavior that contributes to the overall molecular response of the ECM. Other than the ECM, it is the relation of cells with neighboring cells that has an impact on their physiological outcome. In other words, the idea that gene expression in cells is regulated and governed less by its metabolic activity and more by physiological and environmental conditions is emphasized. Cells dynamically interact with their surroundings which dictate their internal machinery to respond. Most often these interactions involve direct cell-cell and/or cell-ECM contacts which differ in their molecular constitutions. The importance of ECM is further acknowledged through the disorders caused by the mutations in ECM related genes.

ECM understanding leads us to comprehend the natural environment that allows the cell to remain healthy and respond in a physiologically distinct manner. Ideas of mimicking the ECM to impart a native tissue-like environment to the cells ex vivo are described in Chapter 3. Different models and materials for culturing cells in a 3D environment that are being explored are also covered. A broad spectrum of contemporary techniques (which may not be all inclusive) for creating 3D substrates from different schools are discussed. Applications of ECM mimicking scaffolds go much beyond tissue engineering (TE) and regenerative medicine (RM). They are briefly mentioned while those in TE & RM are elaborated on.

The final chapter includes available technologies for specific purposes with relevant technical details. The wide range of approaches and techniques already being explored with different targeted objectives could provide an enlightening perspective to readers. It also presents the expanded horizon and scope for improvements, which is expected to encourage others to come up with fresh ideas and pursue them to contribute to the field.

Chapter 1

Introduction

1.1 Cell Culture: Historical Perspective

Nature's ways of producing diverse kinds of cells while imparting specific functionality to different tissues has been quite enigmatic. There has always been a fine line between healthy (normal) and unhealthy (abnormal) tissue. In order to keep a tissue/organ functionally active and healthy not only the cells, which are actively involved in functional manifestations, but also those which co-ordinate to keep them active in that manner are equally important. These supporting cells, often referred to as stromal cells, take care of the metabolic output of a tissue/organ through signal balancing. The complex macromolecular network surrounding the cells also plays a vital role in dispensing the appropriate signals to the cells. Hence, the health of a tissue though largely depends upon the cells directly involved in originating the functionality; the surrounding cells and extracellular microenvironment consistently influence their optimal behavior. The contribution of supporting cells and extracellular matrix (ECM) is often remarkable in keeping the functional integrity of cells/tissue intact [1].

Cells are known to multiply through mitosis occurring at regular intervals. They also grow in size and differentiate into various

3D Cell Culture: Fundamentals and Applications in Tissue Engineering and Regenerative Medicine
Ranjna C. Dutta and Aroop K. Dutta
Copyright © 2018 Pan Stanford Publishing Pte. Ltd.
ISBN 978-981-4774-53-6 (Hardcover), 978-1-315-14682-9 (eBook)
www.panstanford.com

subtypes to perform different kinds of functions. Microscopic inspections help in recording morphological differences, if any, between healthy and diseased cells. However, it does not decipher the extent of functional malignancy in diseased cells. Therefore, to gain a better insight into cellular behavior the need of growing cells ex vivo was felt. Cells isolated from both plant and animal sources could successfully be kept alive by providing glucose and other essential nutrients in a culture flask or on a Petri dish normally used for culturing bacteria. Medium containing growth factors, vitamins and other energy supplements with specified temperature and pH buffering are used for optimized results. However, unlike micro-organisms animal cells have limited cell division capacity under artificial conditions and it was found difficult to follow animal cells for a longer period of time. To overcome this inadequacy cancerous cell lines with unlimited cell division capacity were developed. This also helped in countering the discrepancy in the results observed due to variation in the source of the cells. Cell line depositories were thus created to keep track and accountability of cell culture outcome. Thereafter in vitro cell culture was accepted as a routine technique in different laboratories the world over. Cell culture methods contributed immensely in bringing our understanding about cellular and molecular biology to the present level. Such techniques also made their way in the chain of drug discovery and diagnostics, where they are now considered almost indispensable. Testing the efficacy and potential of prospective drug molecules and also evaluating their toxicity on relevant cells in vitro became an essential feature of the drug discovery process. Depending upon the availability of appropriate cell lines, the tentative drug is preferred to be evaluated for their pharmacological influence in vitro by cell culture techniques before undertaking animal studies. Drugs designed and developed for killing cancer cells are also screened using cell culture methods. Such techniques are also adapted in cancer diagnosis and testing the sensitivity of malignant tissue towards anticancer drugs before embarking upon the treatment regime for the patient.

Adherent cells are normally grown in a Petri dish or T-flasks where they tend to adhere to the surface and multiply to form a monolayer. Once confluent they stop multiplying and rather start dying. With no more surfaces to adhere and in the absence of appropriate cell-cell connections further growth and functional differentiation is restricted. Non-adherent cells are cultured in roller bottles where cells are suspended in the growth medium and rotated slowly. However, both the methods fall short of providing adequate conditions for the cells to grow and functionally replicate their counterparts in vivo. These procedures though inappropriate for functional replication of cells have been used successfully for recombinant protein and vaccine production. Production of such bio-chemicals utilizes only one aspect of cell function, i.e., the protein expression machinery. Specific antigenic proteins used for vaccination could therefore be expressed through genetically modified cells that can be cultured in roller bottles. However, the yield is quite compromised due to the culture conditions which vary from case to case and need optimization each time. Nevertheless, when it comes to comprehending the actual behavior of specific cells or their response to a particular drug or stimuli we cannot expect the factual picture by growing cells as monolayer in isolation.

1.2 Cell Culture: 2D vs. 3D

The history of cell culture techniques dates back to the late nineteenth century. It was perhaps initiated by Wilhelm Roux who for the first time could maintain the neural plate of chick embryo in saline for a few days [2]. The quest to grow cells artificially and reproduce live tissue in vitro grew slowly with time and in 1907, the beginning of the twentieth century Ross Granville Harrison published an article on culturing nerve cells and monitoring the development of fibers [3, 4]; while Alexis Carrel also attempted to culture heart in 1912 [5, 6]. Using a small fragment of the heart of an 18-day old chick embryo, he demonstrated that over a period of three months with 18 passages the cells remain not only viable but continued their rhythmic beating. This established for the first time that tissues could be cultured in vitro while retaining their normal functions [7]. Leighton in the 1950s introduced sponges made up of gelatin, cellulose and collagen treated cellulose for supporting and maintaining the native tissue-like architecture [8,

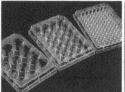

Figure 1.1 Flat-bottomed (2D) tools/appliances conventionally used for cell culture.

9]. He also developed a histophysiologic gradient culture method where reportedly more realistic tissue formation could take place [10]. Although reinforced through the work of Holtfreter [11] and Moscona [12–14] that new insights into the tissue morphogenesis can be achieved by generating the spherical aggregates of isolated embryonic cells, the next two decades evidenced little progress in this direction. The 3D tissue culture approach was hijacked by the convenience of dish culture methods, which was originally introduced for prokaryotic cell culture. Petri dishes first introduced by Petri to grow prokaryotes [15] were hereafter indiscriminately adapted for growing eukaryotic cells as well (Fig. 1.1).

Even though time-honored, conventional cell culture in a Petri dish provides unnatural conditions with only 2D space for growing cells. Thus, in spite of initial steady development through cultivating cells in their natural, native resembling, three dimensional (3D) environments, we lost track. That surface flattening could result in physiologically compromised cells is now an established fact. Tissue/cells in the body grow in three dimensions surrounded by other cells and an extracellular matrix (ECM) bathed in blood plasma or tissue fluid. The flat and rigid plastic/glass surface and culture medium supplemented with fetal wound fluid (fetal bovine serum or FBS) provide an environment completely alien to the normal cells [16]. Blood serum, unlike plasma, consists of several factors that trigger fibroblast multiplication and divert tissue homeostasis from the normal functionally differentiated path to a typical wound healing response. Consequently, the fibroblast subpopulation multiplies quickly to outnumber other specialized functional cells, leading to impairment of specialized tissue functions. Though in conventional cell culture various growth factors and other important constituents are supplied in media, yet in the absence of a natural matrix-like environment where many such proteins are presented in a bound state, the normal growth and differentiation is mislaid. Glass and plastic Petri dishes are easy-to-handle but are undeniably compromised on the microenvironment existing in vivo, thereby estranging our findings away from the real [17]. It is now agreed that the idea of cellular evaluation in three-dimensional spaces was ignored presumably in the face of convenience offered by the two-dimension (2D) culture possible on flat-bottomed plates and T-flasks. Eventually, our perception on both physiologically fit (healthy) and unfit (pathological) cells at present has been by and large created and established by the flat culture systems. Even co-culture with accomplice cells was rare. Though in many cases these conditions successfully reproduced morphologically identical cells, yet only in a few occasions could the physiological uniqueness of the native cell type be maintained/preserved. In a majority of cases cells lost their physiological behavior after a few passages ultimately leading to the predominance of fibroblast-like cells.

Realizing the importance of the microenvironment for retaining normal cell behavior Elsdale & Bard in 1972 questioned the ongoing cell-culture practices through a statement "*Snatched from a life of obscurity and installed in contemporary glass and plastic palaces, cells are in danger of becoming Pygmalion's protégés." Housed in more traditional residences constructed of water and collagen instead of plastic or glass, do cells lead primitive, less cultured lives* [18]. Surely it was a valid question considering the transformation from tissue culture to monolayer culture where single type of cells are grown in isolation on a Petri dish while ignoring the requisite native three dimensional microenvironment. By this time using flat-bottomed surfaces with isolated cell lines became the norm not only for testing the drug efficacy and sensitivity but also for diagnostic assays. The Petri dish culture gave us convenience but compromised with the real native-like behavior of cells. No wonder this convenience contributed to the failure rate of new drug entities which have been tested in vitro using similar systems throughout. The chances of severe impact is very high as the cells, be normal or cancerous, grown in 2D are not the true representative of cells in vivo. It took time to re-establish the importance of a third dimension and biological mimicking to discern the true picture of cell behavior [7]. The very fact that cells, which normally co-exist with other cells in a 3-dimensional space in vivo deviate physiologically when restricted to grow in two dimensions is acknowledged and highlighted all over again [19–21]. Biologists started revisiting Carrels and Leightons' procedures of growing cells in sponges and collagen matrices. Critical comparisons were attempted between two-dimensional (2D) monolayer culture and the culture done on three-dimensional (3D) spheroids [22]. Spheroids are the cell-aggregates generated by growing cells in spinner flasks, a technique developed in 1970 by Sutherland and his coworkers [23]. Cells appear morphologically very different in 2D vs. spheroid (Fig. 1.2).

Heppner and his colleagues in the context of cancer demonstrated for the first time that the tissue architecture plays a critical role in determining the sensitivity towards a particular drug [24, 25]. Obviously it had a great impact on the way drug molecules should be screened vs. the ongoing practice. Spheroids resembled the tumors only morphologically and were therefore the most simplified representation, yet they provided better accuracy in terms of drug efficacy. Such studies have drawn the attention and helped in establishing the importance of the extracellular microenvironment (ECM) for more accurate cell response. They also suggested that providing a third dimension to the growing cells could restore and maintain the differentiated status of adult cells in vitro. Reports comprehending the importance of proteins, glycoproteins, proteoglycans and other macromolecules present in the extracellular space in modulating cellular response started appearing in literature. Collagen, the major ECM protein, was accounted for promoting the reorganization of the pancreatic endocrine cell monolayer into islet-like organoids [26].

Advances in Development Biology in the meantime also suggested that the contribution of ECM proteins goes much beyond their architecture. It was discovered that during embryo development ECM plays an important role in regulating the growth and differentiation for steering it in the right direction [27, 28]. Embryonic development studies further disclosed that structural integrity of ECM is crucial not only after but even before fertilization [29, 30]. Value of the third dimension was also appreciated

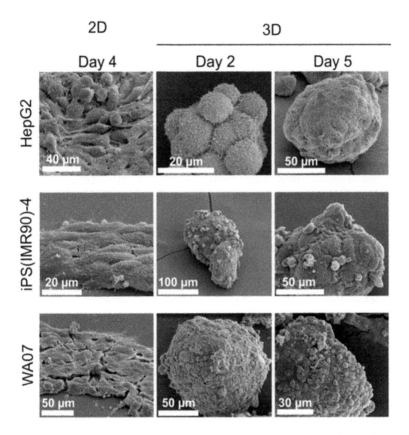

Figure 1.2 Morphology of human pluripotent stem cells iPS(IMR90)-4, WA07 and hepatocellular carcinoma (HepG2) cells grown on flat surface (2D) and as spheroid (3D) on plant-derived nanofibrillar cellulose (NFC) hydrogel. (3D spheroids are preserved using silica bioreplication method.) Adapted with permission from Lou et al., *Sci Rep*, 2015, **5**, article number 13635; doi:10.1038/srep13635.

while following the growth and differentiation pattern of different progenitor cells. Most of these cells are found to require at least temporarily, an aggregation into embryoid bodies for proper differentiation [28]. Multiple types of cells growing together are observed to help each other in acquiring and retaining organ like functionality. Similar cells growing cohesively are found to adapt supportive positions to co-ordinate and manage the tissue specific function of the core group of cells. Hepatocyte spheroids for example are found to be physiologically more expressive and tend to manifest liver-like features [31–33]. Thus, besides cell-ECM, cell-cell interaction and communication is also important for acquiring tissue like functionality. This illustrated the significance of cellular organization and ECM in the development, differentiation and functional efficiency of a tissue [34]. Cells signal each other through ECM modulations, which often require matrix metalloproteinase activity.

Use of chemically and architecturally different 3D culture designs boosted the research on developing multipotent, hematopoietic progenitors [35, 36]. Chondrocytes are found to re-express cartilage specific genes when cultured in alginate beads [37, 38] and osteoblasts cultured in three dimensions for the first time showed proliferation and differentiation potential in vitro [39, 40]. In vitro cartilage tissue engineering is also made possible by the 3D culture of mesenchymal stem cells [41, 42]. As expected, culturing cells in 3D brought in newer parameters, which were never taken into account during conventional monolayer culture. It has since been realized that some of the spatial and temporal features of the extracellular matrix also play an important role in presenting and maintaining the gradients of soluble factors (growth factors, cytokines and hormones) [43]. This resulted in generating realistic information much closer to that existing in a natural, complex set up [44]. Unlike monolayer culture on traditional smooth surfaces, the cell-cell [45] and cell-surface interactions are also altered in cells growing in 3D [46].

Thus, it is now established that imparting freedom to expand in 3-dimensional spaces allows the cells to acquire natural characteristics and resemble much more closely to cells growing in vivo. However, the physical and chemical nature of 3D material support has its influence in determining the fate of the cells. This impact is mainly through the surface topography [47] and other molecules and functionalities of the matrix which are available for direct cell interaction [48]. Such interactions ultimately regulate the cell-cell and cell-ECM dynamics. Along with this, the mechanical force like the fluid flow within the matrix is also important for the health of the growing cells [47, 49].

Before proceeding to the factors important for consideration to design a scaffold for 3D culture, we need to understand the fundamentals of the extracellular microenvironment and the underlying mechanisms of cell-signaling in brief. Reflecting the relevance of the extracellular matrix (ECM) and its tissue specificity in the context of the 3D scaffold design, the next chapter is devoted to important components of the extracellular matrix and their perceptive role in the overall grooming of cells.

References

1. Caspi O, Lesman A, Basevitch Y, Gepstein A, Arbel G, Habib IH, Gepstein L and Levenberg S (2007). Tissue engineering of vascularized cardiac muscle from human embryonic stem cells. *Circ Res*, **100**, 263–272.

2. Rodríguez-Hernández CO, Torres-García SE, Olvera-Sandoval C, Ramírez-Castillo FY, Loera Muro A, Avelar-Gonzalez FJ and Guerrero-Barrera AL (2014). Cell culture: history, development and prospects. *Int J Curr Res Aca Rev*, **2**(12), 188–200.

3. Harrison R (1907). Observations on the living developing nerve fiber. *Anat Rec*, **1**, 116–128.

4. Harrison R (1910). The outgrowth of the nerve fiber as a mode of protoplasmic movement. *Proc Soc Exp Med*, *NY*, 140–143.

5. Carrrel A (1912). On the permanent life of tissue outside the organism. *J Exp Med*, **15**(5), 516–528.

6. Carrel A and Lindberg C (1938). The culture of organs. *Can Med Assoc J*, New York: Paul B. Hoeber, Inc., Medical Book Department of Harper & Brothers.

7. Hoffman RM (1993). To do tissue culture in two or three dimensions? That is the question. *Stem Cells*, **11**, 105–111.

8. Leighton J (1951). A sponge matrix method for tissue culture; formation of organized aggregates of cells in vitro. *J Natl Cancer Inst*, **12**(3), 545–561.

9. Leighton J, Justh G, Esper M and Kronenthal R (1967). Collagen-coated cellulose sponge: three-dimensional matrix for tissue culture of Walker tumor 256. *Science*, **155**(3767), 1259–1261.

10. Leighton J (1992). Structural biology of epithelial tissue in histophysiologic gradient culture. *In Vitro Cell Dev Biol*, **28A**(7–8), 482–492.

11. Holtfreter J (1944). A study of the mechanics of gastrulation: Part II. *J Exp Zool*, **95**(2), 171–212.

12. Moscona A (1952). Cell suspensions from organ rudiments of chick embryos. *Exp Cell Res*, **3**, 535–539.

13. Moscona A (1957). The development in vitro of chimeric aggregates of dissociated embryonic chick and mouse cells. *Proc Natl Acad Sci USA*, **43**(1), 184–194.

14. Moscona A (1961). Rotation-mediated histogenetic aggregation of dissociated cells. A quantifiable approach to cell interactions in vitro. *Exp Cell Res*, **22**, 455–475.

15. Petri RJ (1887). Eine kleine Modification des Kochśchen Plattenverfahrens. *Centralbl Bacteriol Parasitenkunde*, **1**, 279–280.

16. Schmeichel KL and Bissell MJ (2003). Modeling tissue-specific signaling and organ function in three dimensions. *J Cell Sci*, **116**(12), 2377–2388.

17. Altmann B, Welle A, Giselbrecht S, Truckenmüller R and Gottwald E (2009). The famous versus the inconvenient - or the dawn and the rise of 3D-culture systems, *World J Stem Cells*, **1**(1), 43–48.

18. Elsdale T and Bard J (1972). Collagen substrata for studies on cell behavior. *J Cell Biol*, **54**(3), 626–637.

19. Li ML, Aggeler J, Farson DA, Hatier C, Hassell J and Bissell MJ (1987). Influence of a reconstituted basement membrane and its components on casein gene expression and secretion in mouse mammary epithelial cells. *Proc Natl Sci USA*, **84**(1), 136–140.

20. Zahir N and Weaver VM (2004). Death in the third dimension: apoptosis regulation and tissue architecture. *Curr Opin Genet Dev*, **14**, 71–80.

21. Abbott A (2003). Cell culture: biology's new dimension. *Nature*, **424**, 870–872.

22. Li LH, Bhuyan BK and Wallace TL (1989). Comparison of cytotoxicity of agents on monolayer and spheroid systems. *Proc Am Assoc Cancer Res*, **30**, 2435a.

23. Inch WR, McCredie JA and Sutherland RM (1970). Growth of nodular carcinomas in rodents compared with multi-cell spheroids in tissue culture. *Growth*, **34**, 271–282.

24. Lawler EM, Miller FR and Heppner GH (1983). Significance of three dimensional growth patterns of mammary tissues in collagen gels. *In Vitro*, **19**(8), 600–610.

25. Miller BE, Miller FR and Heppner GH (1985). Factors affecting growth and drug sensitivity of mouse mammary tumor lines in collagen gel cultures. *Cancer Res*, **45**, 4200–4205.

26. Montesano R, Mouron P, Amherdt M and Orci L (1983). Collagen matrix promotes reorganization of pancreatic endocrine cell monolayers into islet-like organoids. *J Cell Biol,* **97**, 935–939.

27. Tsang KY, Cheung MC, Chan D and Cheah KS (2010). The developmental roles of the extracellular matrix: beyond structure to regulation. *Cell Tissue Res,* **339**, 93–110.

28. Frantz C, Stewart KM and Weaver VM (2010). The extracellular matrix at a glance. *J Cell Sci,* **123**, 4195–4200.

29. Monné M, Han L, Schwend T, Burendahl S and Jovine L (2008). Crystal structure of the ZP-N domain of ZP3 reveals the core fold of animal egg coats. *Nature,* **456**(7222), 653–657.

30. Wassarman PM, Jovine L and Litscher ES (2004). Mouse zona pellucida genes and glycoproteins. *Cytogenet Genome Res,* **105**(2–4), 228–234.

31. Koide N, Shinji T, Tanabe T, Asano K, Kawaguchi M, Sakaguchi K, Koide Y, Mori M and Tsuji T (1989). Continued high albumin production by multicellular spheroids of adult rat hepatocytes formed in the presence of liver-derived proteoglycans. *Biochem Biophys Res Commun,* **161**, 385–391.

32. Wu FJ, Friend JR, Remmel RP, Cerra FB and Hu WS (1999). Enhanced cytochrome P450 IA1 activity of self-assembled rat hepatocyte spheroids. *Cell Transplant,* **8**, 233–246.

33. Abu-Absi SF, Friend JR, Hansen LK and Hu WS (2002). Structural polarity and functional bile canaliculi in rat hepatocyte spheroids. *Exp Cell Res,* **274**, 56–67.

34. Adams JC and Watt FM (1993). Regulation of development and differentiation by the extracellular matrix. *Development,* **117**, 1183–1198.

35. Bagley J, Rosenzweig M, Marks DF and Pykett MJ (1999). Extended culture of multipotent hematopoietic progenitors without cytokine augmentation in a novel three-dimensional device. *Exp Hematol,* **27**, 496–504.

36. Evans MJ and Kaufman MH (1981). Establishment in culture of pluripotential cells from mouse embryos. *Nature,* **292**, 154–156.

37. Bonaventure J, Kadhom N, Cohen-Solal L, Ng KH, Bourguignon J, Lasselin C and Freisinger P (1994). Reexpression of cartilage-specific genes by dedifferentiated human articular chondrocytes cultured in alginate beads. *Exp Cell Res,* **212**, 97–104.

38. Guo JF, Jourdian GW and MacCallum DK (1989). Culture and growth characteristics of chondrocytes encapsulated in alginate beads. *Connect Tissue Res*, **19**, 277–297.

39. Ferrera D, Poggi S, Biassoni C, Dickson GR, Astigiano S, Barbieri O, Favre A, Franzi AT, Strangio A, Federici A and Manduca P (2002). Three-dimensional cultures of normal human osteoblasts: proliferation and differentiation potential in vitro and upon ectopic implantation in nude mice. *Bone*, **30**, 718–725.

40. Granet C, Laroche N, Vico L, Alexandre C and Lafage-Proust MH (1998). Rotating-wall vessels, promising bioreactors for osteoblastic cell culture: comparison with other 3D conditions. *Med Biol Eng Comput*, **36**, 513–519.

41. Wang Y, Kim UJ, Blasioli DJ, Kim HJ and Kaplan D (2005). In vitro cartilage tissue engineering with 3D porous aqueous derived silk scaffolds and mesenchymal stem cells. *Biomaterials*, **26**, 7082–7094.

42. Grayson WL, Ma T and Bunnell B (2004). Human mesenchymal stem cells tissue development in 3D PET matrices. *Biotechnol Prog*, **20**, 905–912.

43. Discher DE, Mooney DJ and Zandstra PW (2009). Growth factors, matrices, and forces combine and control stem cells. *Science*, **324**, 1673–1677.

44. Pampaloni F, Reynaud EG and Stelzer EHK (2007). The third dimension bridges the gap between cell culture and live tissue. *Nat Rev Mol Cell Biol*, **8**, 839–845.

45. Yamada KM, Pankov R and Cukierman E (2003). Dimensions and dynamics in integrin function. *Braz J Med Biol Res*, **36**, 959–966.

46. Cukierman E, Pankov R and Yamada KM (2002). Cell interactions with three-dimensional matrices. *Curr Opin Cell Biol*, **14**, 633–639.

47. Bettinger CJ, Langer R and Borenstein JT (2009). Engineering substrate topography at the micro- and nanoscale to control cell function. *Angew Chem Int Ed Engl*, **48**, 5406–5415.

48. Schwartz MA and De Simone DW (2008). Cell adhesion receptors in mechanotransduction. *Curr Opin Cell Biol*, **20**, 551–556.

49. Pedersen JA and Swartz MA (2005). Mechanobiology in the third dimension. *Ann Biomed Eng*, **33**, 1469–1490.

Chapter 2

Significance of Tissue-Specific Extracellular Microenvironment (ECM)

2.1 Introduction

Cells and their nuclei are the smallest functional units that work in tandem in a living organism. A functional tissue/organ on the other hand incorporates a number of synchronized cells committed in concert to execute certain physiological tasks. Carrying huge, organism-specific genetic information compacted in their nucleus, cells co-ordinate with neighboring cells to impart specific functionality to a tissue. Except in blood and lymph the tissue-specific cells are confined to a well defined region where they network through the extracellular microenvironment/matrix (ECM) to accomplish a specific metabolic output. Cells receive and tend to respond to a variety of signals/stimuli presented through their immediate microenvironment. These signals come into effect either through the binding of fresh molecules or as a result of the conformational transition of different domains of ECM constituents that could alter the pre-existing cell-ECM interaction. Structural modulation of ECM constituents like proteoglycans, glycoproteins is often a consequence of the functional interaction with other

3D Cell Culture: Fundamentals and Applications in Tissue Engineering and Regenerative Medicine
Ranjna C. Dutta and Aroop K. Dutta
Copyright © 2018 Pan Stanford Publishing Pte. Ltd.
ISBN 978-981-4774-53-6 (Hardcover), 978-1-315-14682-9 (eBook)
www.panstanford.com

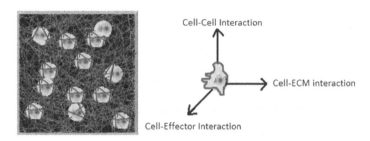

Figure 2.1 Schematic of cells imbedded in the extracellular matrix (ECM) network.

smaller effectors like growth factors, cytokines, hormones, etc. (Figs. 2.1 and 2.2 for schematic view). Cells of one tissue type have to depend on connective tissues to communicate with cells of other tissue/organ type. These signals could be passed from adjacent cells via focal adhesions [1] or may approach as soluble factors (hormones) brought about through blood or lymphatic fluid from distant organs, or spatially presented by means of localized modulation of ECM. The signaling cues could thus reach to other tissues/organs through three different modes; autocrine (self-signaling), paracrine (vicinity/adjacent signaling) and endocrine (distant signaling through connective tissues). In a tissue where cells have to co-ordinate to yield a coherent physiological function, communication happens directly via cell-cell junctions and/or gated channels as well. Direct communication of this kind helps the cells to acquire a polarized state, which is essential to impart a defined shape to a tissue. Polarized arrangement of cells and the intercellular tight junctions also contribute to the osmotic and barrier function of the consequently formed cellular layer.

The fact that lineage is not the only factor that guides or controls cell response is already established through developmental biology. Despite sharing identical genome, cell phenotypes from an embryo turn out to be different [2]. The blastula, a ball of cells formed after preliminary cleavage of fertilized egg, if not placed in an appropriate microenvironment fails to follow the accurate and requisite differentiation path. As a consequence it never delivers the physiologically viable organism that it sets to transform. Continuous contact and communication with the neighboring cells also contribute equally for successful transition to a physiologically functional organism. Cell-cell communication plays an important role throughout the gestation period and could regulate and deregulate tissue specific differentiation. Gestational programing exhibit the survival instinct to an appreciable extent through nutritional tolerance and gets prolonged or shortened accordingly.

In general the biopolymers and functional macromolecules that surround different cells and span in three dimensions of a functional tissue constitutes its ECM. ECM remodels dynamically under the influence of occupying cells in a tissue. Traditionally thought to serve only by providing support and strength to cells, ECM is now recognized to play a much bigger role during development and growth. It is now established that ECM is the major contributor in regulating morphogenesis and organogenesis. Cell processes like adhesion, proliferation, migration, survival and differentiation could be altered through ECM modulation [3, 4]. Intercellular signaling which is mediated via ECM may trigger positive or negative feedback mechanisms in the cells.

Different animal models ranging from small worms (*Caenorhabditis elegans*), fruit flies (various *Drosophilia* species), zebrafish (*Danio rerio*) to frogs (*Xenopus laevis* and *Xenopus tropicalis*), chicken (*Gallus gallus*), and mouse (*Mus musculus*) have been used to establish the importance of ECM. However, among them fruit flies having short life spans, a smaller genome linked with distinct morphological features are the most preferred to study the genetic alteration and correlation study [3, 4]. Embryo development studies in chicks also helped in exploring the role of ECM in cell migration and differentiation in the early embryonic stages. Gene knock-out models too find use in identifying and establishing the magnitude of implication of ECM constituents in the survival of an organism.

2.1.1 What Is ECM?

By definition the extracellular microenvironment/matrix (ECM) represents the surroundings of cells/tissues, which primarily incorporate nonliving macromolecules, including insoluble and soluble protein fibrils, polysaccharides, glycoproteins and proteoglycans in

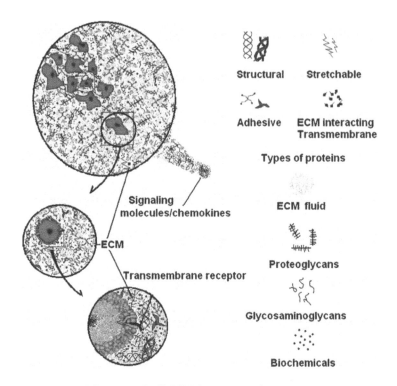

Figure 2.2 Schematic of cell-ECM dynamics and its major participants.

the extracellular space [3, 5–8]. The ECM is an organized meshwork of macromolecules in close association with the cells that produce them. The ECM plays multiple important roles, among them the most important is being an anchorage to the cells (Fig. 2.1). Besides providing structural support to the organized physiologically active cellular structures called tissue or organ, it also mediates the information exchange among them. As more complex animals have evolved, the ability of ECM to signal for the vasculature and its remodeling also evolved. Particularly in mammals, evolution enabled the ECM to signal the cells for initiating or terminating the processes involved in blood vessel formation [9]. This ability is crucial in view of the need to supply nutrition and oxygen to the organism's extremities for cellular survival.

Often referred to as connective tissue, ECM also helps in joining different tissues and organs of the body together. Connective tissue, other than blood and lymph, incorporates only a few cells imbedded into fibrous but elastic ECM. Fibrous proteins like collagen and elastin and hydrophilic molecules like proteoglycans and glycoproteins are the major constituents of such ECM. The structure and in turn the function of ECM depends on the relative proportions of these constituent molecules. Tissues that have to endure tensional forces to remain stretched like tendons are rich in fibrillar collagens, while the ones required to withstand compressive forces such as cartilage possess high content of proteoglycans [7].

The ECM adjacent or peripheral to cells is sometime referred to as pericellular matrix or PCM [10]. The macromolecules in this region are mostly entangled or integrated with the cell membrane at least through one of their end terminus. The composition of the pericellular matrix (PCM) may also be distinctly different from the extracellular matrix. In articular cartilage, for example collagen VI fibrils are found confined to only the PCM region although aggrecan and collagen II are common in both PCM and ECM [11, 12]. Acting as a bridge between ECM and cell, PCM acts as a second control that filters and regulates the soluble factors like cytokines, growth factors, matrix catabolites, etc., again before they finally interact with the cells [13]. Therefore, extracellular cues that affect the development and/or maintenance of the PCM may in turn directly impact the signals perceived by the cells and subsequently affect the long-term formation and/or maintenance of the ECM [14].

The ECM molecular network is crucial in providing both structural and mechanical integrity to the tissues. This pericellular and extracellular network also functions as a depository for growth factors, hormones and other chemotactic signaling agents and provides a medium through which nutrients and chemical messengers can diffuse. They mediate the physicochemical environment and also make water and other ions accessible to the cells. Mechanical and biochemical changes in the ECM region are also responsible for cell mobility within the connective tissue scaffold. The fibrillar yet viscoelastic nature of ECM helps in sustaining the tissue architecture during stressful activities. It also plays a key role in the normal performance of a group of cells by assisting the coordination among the cells for specified tissue function. Cell movement and migration through adhesive molecules that recognize cell surface receptors is also facilitated through ECM. The constituting extra cellular macromolecules may be similar in different tissues but their relative composition and morphological organization is always specific to that particular tissue. ECM proteins in general consist of structurally independent domains whose sequences and arrangement are highly conserved. These domains mediate cell-matrix adhesion and act as transducers of signals. ECM proteins also bind to soluble growth factors and regulate their distribution, activation, and presentation to cells [8]. As an organized, solid-phase platform an ECM protein is capable of integrating complex, multivalent signals to cells in a spatially patterned and regulated manner. A number of molecules influence the architecture and physiological efficiency of ECM [15]. Vitamins, hormones, growth factors, matrix metalloproteinases (MMPs) and their inhibitors are some of which known to have a direct impact on the ECM production. The role of vitamin C and D respectively in the matrix of skin and bone and the importance of growth factor TGF-β in the ECM production and degradation are already established. The relationship of growth factors with the matrix involves a complex set of feedback loops and ligand/storage functions. The growth factors, which modulate matrix structure and synthesis, remain in a bound state on different matrix proteins. They are released following exposure to proteases or other molecular signals. For example fibroblast growth factor, owing to its strong affinity remains attached to heparin. It is released on exposure to proteases such as thrombin or macrophage derived plasminogen activators. Thus, ECM functions as an active modulator of growth factor bioavailability and signaling [16, 17]. Chemically and architecturally optimum ECM is crucial for the efficient and orchestrated functioning of different types of cells of a tissue, which in turn influences the overall functional outcome of that particular organ. In some organs like the kidneys where exchange of nutrients and wastes takes place, the ECM network has a very specialized function to carry out. Similarly in lungs also, the ECM contributes to their unique role of air exchange. The efficient functioning of these organs depends on the distinctive filtration capacity of their ECM network.

ECM exists in a dynamic state, continually being remodeled by the cells, which tend to reshape it as part of their metabolic activity. ECM remodeling is at the core of differentiation as has already been demonstrated by the finding that unlike adipocytes from wild-type mice that differentiate in a 3D collagen I matrix, the MT1-matrix metalloproteinase (MMP) null adipocytes failed to differentiate in MT1-MMP null mice [18]. ECM-remodeling rates are particularly high during development and wound repair, in response to infectious agents and in many disease states [19]. ECM remodeling is induced by its direct interaction with cellular receptors like integrins and also the enzymes like matrix metalloproteinases (MMPs) and other extracellular proteases [20–23]. Mechanical forces and cellular tension can also remodel the ECM through exposure of encrypted signaling moieties. ECM remodeling is also observed under conditions like normal ageing, wound and tumor formation. Precise cause and effect details of this dynamic relation between cells and its ECM are, however, not yet clear [24].

ECM responses are found to be sensitive even to the gradient and differential signals. Recent studies have shown that the cells involved in migration during development and those in metastasis operate under the influence of different growth factor induced signaling pathways [25]. Furthermore, it is observed that in a group of migrating cells precision is achieved by reshuffling the leader among the cells to the one closest to the ligand source and sensing it the best [25]. This suggests that it might be possible to distinguish between the cancerous migrations from the benign ones that occur during development.

Cells adhere and interact with their ECM environment mostly through a class of surface receptor proteins called integrins. Integrins are cell membrane proteins, which connect the ECM with cytoplasmic molecules and the actin cytoskeleton. Such interactions are decisive in activating a variety of intracellular signaling pathways. Integrins are also crucial for transmitting many extracellular signals in stem cells as well [26, 27]. As one of the major adhesion receptors, integrins can signal in both directions across the plasma membrane [28]. Cadherins, selectins, tetraspanins, immunoglobulin receptors and mucins are other classes of proteins that represent cell surface adhesion molecules. It is the unique character of the adhesion complexes formed between these adhesion molecules on the cells and their ECM which brings specificity and triggers a response through a particular downstream signaling pathway in the cell [29].

2.1.2 Tissue Specificity of ECM

In the vertebrate body, ECM exists in four distinct phases (Table 2.1). In bones, cartilage and musculoskeletal tissue it is in a solid state; in blood and lymph it is in a free flowing fluid state while in rest of the other organs it exists in the semisolid gel-like state. Bone and cartilage ECM consists of mainly fibrillar collagens blended with hydroxyapatite (HaP/HAP) like inorganic material and other minerals that impart the much needed strength to the tissue/organ. The semi-solid and fluid phase ECM is interchangeable and differs only in their fluidic and elastic behavior. In other words the stoichiometric ratio of fibrous versus soluble constituents, in general, can dictate the physical state of it. Basement membrane (BM) represents the fourth unique type of ECM which is in sheet format. BM is a ubiquitously present membranous ECM (0.1 to 2 micron thick). It separates epithelial and endothelial cells from underlying

Table 2.1   Types/phases of extracellular matrix

Solid phase as in bones and cartilage
Lamellar components (hydroxyapatite and inorganic salts predominate)
Fibrous components (collagens)

Semi-solid/gel phase as in brain and lungs
(also as filler in interstitial tissue space)
Fibrous components (collagens and proteoglycans)
Soluble components (glycoproteins, matrikines and matricryptins)

Fluid phase as in blood and lymph
(also as filler in interstitial tissue space)
Soluble components (pro-collagens, PGs and small functional molecules like fibronectin and matrikines)
Viscous components (glycosaminoglycans and hyaluronan)

Basement membrane as in skin and vascular linings
Lamellar components (laminins predominate)
Fibrous components (collagen IV predominates)

---

mesenchyme in the tissues. It is present at the interface to demarcate different tissues from neighbouring connective tissues. Present as a homogenous sheet of extracellular material basement membrane is devoid of cells. Type IV collagen, laminin, nidogen/entactin, and heparan sulfate proteoglycan perlecan are the four major components of BM [30, 31]. Thus various ECM components and their stoichiometric ratio are grossly responsible for the variation in the physical state of ECM in respective tissues. They are able to impart a defined shape, structure, elasticity, flexibility and also a range of adhesive and anti-adhesive properties to the respective tissues/organs. Architecturally, tissues may be ECM intensive or cell intensive, which apparently is governed by their physiological relevance and need. For example the ECM component in connective tissue like blood and cartilage is in abundance whereas heart and kidney-like tissues are predominated by the embedded cells. In such tissues, physiologically specialized cells are cohesively arranged to take a compact shape which is invariably maintained by a barrier membrane anteriorly lined with BM.

ECM molecules are remarkably diverse in their structure and distribution. Table 2.2 shows a broad classification of extracellular matrix macromolecules. Protein and glycoprotein constituents of ECM in different tissues differ in their chemical compositions as well as relative proportion. ECM is therefore tissue specific and the characteristic features of its major components determine the physical attributes of that tissue. ECM evolution in metazoans demonstrate very high conservation of core ECM design and that the ECM components do not function in isolation but depend on posttranslational processing enzymes, cell surface receptors, and extracellular proteases [32]. In the dynamic relation that exists between cell and its surroundings, most often it is the ECM that is responsible for keeping the cells functionally viable, healthy and differentiated through quick structural adaptations. This has now been witnessed in vitro by monitoring cells in a 2D and 3D environment and also by lineage-transformations that can be enabled simply by changing the cocktail of the medium. Cells can differentiate, de-differentiate and even trans-differentiate (enter a different development pathway) if provided with an appropriate signal [33, 34].

Table 2.2 Extracellular matrix (ECM) components (broad classification)

---

(1) Collagens (fibrous)

   (i) Structural
   (ii) Elastic
   (iii) Adhesive/antiadhesive

(2) Proteoglycans

   (i) Small leucine-rich (SLRP), e.g., decorin, biglycan, lumican, epiphycan
   (ii) Modular

      (a) Nonhyaluronan binding, e.g., perlecan, agrin, testican
      (b) Hyaluronan and lectin binding, e.g., aggrecan, versican, neurocan and brevican

   (iii) Cell-surface/transmembrane (TMP)

      (a) Syndecans
      (b) Glypicans

(3) Glycoproteins

   (i) Interstitial connectors, e.g., fibronectins, tenacins, fibrillins, microfibril-associated link proteins, elastin, fibromodulin, matrillins, thrombospondins and osteopontin
   (ii) Basement membrane associated, e.g., laminins, nidogens, enactin and fibulin

(4) Glycosaminoglycans

   (i) Integrated as proteoglycans, e.g., chondroitin sulfate, heparin sulfate dermatan sulfate
   (ii) Free state, e.g., hyaluronic acid (non-sulfated GAG)

(5) Soluble factors
Micro- and macro-sized ECM components including those in pro-state existing in non-fibril, soluble state, growth factors, hormones, vitamins that constitute the tissue fluid filled in the interstitial space

---

2.1.3 Essential/Representative Components of ECM

The inventory of ECM proteins discovered and identified from various tissues is constantly growing. Morphologically ECM components can be represented as fibrillar, globular, and stratified having a tendency to form matrices. Their combinations may acquire fluid, liquid crystalline, gelly or elastic solid states depending upon the physiological functions of a tissue. Table 2.3, though it may

Table 2.3 Major constituents of extracellular matrix (ECM)

ECM components	Important aspects	Protein size (kDa)	Ref.
Structural proteins	*Nature/class*		
Collagen I-III, V, XI, XXIV & XXVII	Fibrillar	310	1, 2
Collagen IV	Polygonal meshwork	525	
Collagen VII	Anchoring fibrils	590	
Collagen VIII & X	Fine fibrillar	200–225	
Collagen IX, XII, XIV, XX, XXI & XXII	FACIT*	225–590	
Collagen XVI & XIX	FACIT Like		
Collagen XV & XVIII	Multiplexin		
Collagen XIII, XVII, XXIII & XXV	Transmembrane		
Collagen XXVI	Testis & ovary specific		3
Stretchable proteins			
Elastin	Elastic (polymeric)	70	4
Fibrillin	Fibrillar	350	5
EMILIN (elastin microfibril interface located protein)		115	6
Adhesive proteins/glycoprotein			
Entactin/nidogen	Globular	120	7
Laminins	Polygonal cross shape	840	8
Fibulin	Pinhead (lobular with rod)	54–175	9
Fibronectin	Multidomain	500	10
Osteocalcin (bone Gla protein)		6	11
Osteopontin (bone phosphoprotein)	D & E rich	44	12
Osteonectin (Ca^{++}-binding protein) (BM-40)	D, E & C rich	32.7	13
Osteogenin	Bone morphogenic	24	14
Bone sialoprotein (BSP)	E & G rich	33.6	15
Tenascins (janusin, cytotactin/ tenascin-C, restrictin/ tenascin-R, tenascin-X)	Multidomain (oligomeric)	320	16
Reelin (gigantic)	Muti-EGF domains	410	17
Tetranectin	Homotetramer	63	18
TRAMP (Y-rich acidic matrix protein/ dermatopontin)	5 disulfide loops	22	19

(*Continued*)

Table 2.3 *(Continued)*

ECM components	Important aspects	Protein size (kDa)	Ref.
Von Willebrand factor	Multimeric	116	20
Vitronectin	Dimeric	50	21
Proteoglycans	*Type of GAG*	*Core protein size (kD)*	
Aggrecan	HA/CS/KS	320	22
GHAB (glial hyaluronate binding protein)	HA	60	23
BEHAB (Brevican/ brain-enriched hyaluronan binding protein)	HA	3.9	24
Hyaluronectin	HA	68	25
Lumican	KS	42	26
Mimecan (osteoglycin)	KS	25	27
Biglycan	CS/DS	38	28
Betaglycan	HS/CS	110	29
Decorin	KS/CS/DS	36.5	30
Epiphycan	DS	46	31
Fibromodulin	KS/CS/DS	60	32
Keratocan	KS	50	33
Glypican	HS	65	34
Syndecan	HS/CS	35	35
Syndecan-4 (ryudocan)	HS/CS	35	36
Perlecan	HS	600	37
Agrin	HS	225	38
Serglycin	CS/DS	11	39
Link protein	HA	40	40
Neurocan	HA/CS	150–260 (220)	41
Phosphacan	HA/CS	400	42
Versican	HA/CS/DS	250	43
Thrombomodulin	CS	90	44
Thrombospondin	HS	450	45
Glycosaminoglycans (GAG)			
Chondroitin sulfate (CS)	As proteoglycan	–	46
Dermatan sulfate (DS or CS-B)	-do-	–	47
Heparan sulfate (HS)/ Heparin	-do-	–	48
Keratan sulfate (KS)	-do-	–	49
Hyaluronan (HA)	Polygonal meshwork	–	50

not be complete, illustrates major ECM constituents and their molecular diversity. Within each type there exist various subtypes with distinct structural and functional features. Some of these proteins, proteoglycans, glycoproteins and glycosaminoglycans that may differ in their subunit structures and/or their permutation-combination have been identified and characterized. It is observed that the variation in their macromolecular arrangement vis-à-vis adjacent or neighbouring molecular units influences their contribution and role as fibrillar, elastic and/or adhesive in the overall matrix.

As informed earlier, the ECM of each tissue is in dynamic relation with the cells it surrounds and keeps remodeling according to their respective activity. The extent of physiological activity of a tissue influences the rate of remodeling of its ECM. Knockout mice models have helped in decoding and establishing the significance of various matrix molecules. Collagen, elastin, laminin, fibronectin and osteopontin are some of the major ECM proteins that are reported to be integrally associated with the ECM of almost all the vital organs [35]. They are phylogenetically conserved biopolymers respectively isolated from muscles, skin, blood and bones. Collagens are the most abundant and widely distributed protein in the body. Elastins and laminins on the other hand are predominant in vascular structures and the basement membrane. Similarly, fibronectin is found in connective tissue, plasma and other body fluids whereas osteopontin is present in bones. In general ECM consists of fibrous protein network embedded with the hydrated gel of different proteoglycans (PGs), glycoproteins (GPs) and glycosaminoglycans (GAGs). The type and ratio of proteins versus polysaccharides depends upon the specific need of that particular tissue. Hydrated gel resists the compressive forces and provides stress endurance, whereas the fibrous protein matrix imparts flexibility and supple-ness besides anchoring the tissue. There exists a unique order and orchestration in the apparently random architecture of the complex ECM network. This distinctive organization is contributed and maintained via the specific structural and biochemical behavior of its ECM constituents. ECM biopolymers may exist in different polymeric forms which differ in their ability to interact with other nearby proteins, proteoglycans and cellular receptors.

Fibrillar collagens and proteoglycans are the predominant biopolymers that remain universally engaged to create the structural matrix of ECM. Embedded in this basic scaffold lies other smaller proteins, glycoproteins, proteoglycans and glycosaminoglycans (GAGs) that may or may not be structurally as important. Collagens are the proteins, which owing to their fibrillar nature, are largely responsible for architectural organization, strength and shape of the tissue. Elastin on the other hand imparts stretching ability to the tissue ECM and makes them functionally flexible while retaining their shape. Laminin, a glycoprotein, is also an important and essential component of basement membrane. Some common and extensively studied glycoproteins, proteoglycans and GAGs belonging to ECM are briefly discussed below. There are other less known, mostly tissue specific ECM molecules that help in interconnecting the major ECM constituents. These are smaller in size and often referred to as matricellular molecules/elements. They do not contribute to the structural integrity of the ECM but are essential for creating bridges and maintaining the requisite mechanics for adequate ECM-Cell dynamics.

The ECM constituting polymers by and large can be divided into following major categories: (i) fibrillar or structural; (ii) elastic; (iii) adhesive (mostly glycoproteins); (iv) matricellular (anti-adhesive); and (v) soluble biopolymers (fillers), which include matrikines and matricryptins. Nevertheless, a partial structure-function overlap is very common among these categories. Interestingly, the ECM macromolecules are found to adapt different shapes if arranged based on their functional domains. They may acquire different geometrical shapes when stabilized for study, e.g., laminin in an arrow-head, collagen IV monomers in sperm shape with a globular non-collagenous head and a long triple helicle tail, and fibronectin in a U or V shape. Similarly most of the proteoglycans adopt a shape of a bottle brush with core protein as the backbone of the compounded molecule. These shapes are conducive to their networking role in the ECM. However, it has to be born in mind that the shapes are representative only, since the molecules are sufficiently flexible within the system and adjust as per the other molecules co-existing in their physiological environment. The amount and type of ions and pH of the surroundings may also influence their overall interactions.

Here some of the extensively studied, representative proteins, proteoglycans and glycoproteins of ECM are introduced. Readers are recommended to refer to detailed reviews on their respective topics for greater insights.

2.1.3.1 Collagens (fibrillar and structural)

Collagens are the most abundant structural constituent of ECM. The term collagen is derived from the Greek word meaning 'to produce glue'. Collagens are the fibrillar, multi-domain proteins, which by virtue of their sequence diversity and abundance contribute as structural proteins (Table 2.4). Each collagen type has polypeptide sequences that are mostly conserved except for certain specific regions. The sequence variation in individual chains on combining with other chains gives rise to structurally different fibrillar networks. They provide a framework through orchestrated alignment of their fibrils, which is essential for acquiring a specific tissue shape. About 28 different types of collagens involving more than 40 distinct polypeptide chains (alpha chains) have been identified in humans only, and nearly 20 other proteins are reported to contain collagen like domains [36]. Each collagen type is present in a distinct preferential combination of subunits. This may be for the sake of physiological stability or to yield the required functionality for productive interactions.

The fibrous nature of collagens is derived from a unique triple helix arrangement corresponding to a number of (Gly–X–Y)n repeats (X and Y frequently being a proline and hydroxyproline) in long peptide chains. The triple helix domains, characteristic of collagens spans to varying lengths with in-between interruptions of relatively smaller non-collagenous (non-triple helix, viz. globular or linear) regions of protein. The collagen superfamily is highly complex and shows a remarkable diversity in molecular organization, tissue distribution and function [37]. The triple helix of collagen may involve homologous or heterologous chains. They associate with other extra cellular macromolecules to give rise to structurally distinct

Table 2.4 Different species/types of collagen and PDB codes of their alpha chains originated from human

Collagen type	α chains [PDB code]	Chain length (Code_Speciecs)
Collagen I	α1(I) [PDB:2LLP]	P02452 (CO1A1_HUMAN)
	α2(I) [COL1A2]	P08123 (CO1A2_HUMAN)
Collagen II	α1(II) [PDB:2FSE]	P02458 (CO2A1_HUMAN)
Collagen III	α1(III) [PDB:2V53]	P02461 (CO3A1_HUMAN)
Collagen IV	α1(IV) [PDB: 1LI1]	P08572 (CO4A2_HUMAN)
	α2(IV) [PDB: 1LI1]	Q01955 (CO4A3_HUMAN)
	α3(IV) [PDB:4HHV]	P53420 (CO4A4_HUMAN)
	α4(IV) [COL4A4]	P29400 (CO4A5_HUMAN)
	α5(IV) [COL4A5]	Q14031 (CO4A6_HUMAN)
	α6(IV) [COL4A6]	
Collagen V	α1(V) [PDB:1A89]	P20908 (CO5A1_HUMAN)
	α2(V) [PDB:1A9A]	P05997 (CO5A2_HUMAN)
	α3(V) [COL5A3]	P25940 (CO5A3_HUMAN)
Collagen VI	α1(VI) [COL6A1]	P12109 (CO6A1_HUMAN)
	α2(VI) [COL6A2]	P12110 (CO6A2_HUMAN)
	α3(VI) [PDB:1KTH]	P12111 (CO6A3_HUMAN)
	α4(VI) [COL6A4]	A2AX52 (CO6A4_MOUSE)
	α5(VI) [COL6A5]	A8TX70 (CO6A5_HUMAN)
Collagen VII	α1(VII) [COL7A1]	Q02388 (CO7A1_HUMAN)
Collagen VIII	α1(VIII) [PDB:1O91]	(*Mus musculus*)
		P27658 (CO8A1_HUMAN)
Collagen IX	α1(IX) [PDB:2UUR]	P20849 (CO9A1_HUMAN)
	α2(IX) [COL9A2]	Q14055 (CO9A2_HUMAN)
	α3(IX) [COL9A3]	Q14050 (CO9A3_HUMAN)
Collagen X	α1(X) [PDB:1GR3]	Q03692 (COAA1_HUMAN)
Collagen XI	α1(XI) [COL11A1]	P12107 (COBA1_HUMAN)
	α2(XI) [COL11A2]	P13942 (COBA2_HUMAN)
	α3(XI) [COL11A3]	
Collagen XII	α1(XII) [COL12A1]	Q99715 (COCA1_HUMAN)
Collagen XIII	α1(XIII) [COL13A1]	Q5TAT6 (CODA1_HUMAN)
Collagen XIV	α1(XIV) [PDB:1B9P]	(*Gallus gallus*)
		Q05707 (COEA1_HUMAN)
Collagen XV	α1(XV) [PDB:3N3F]	P39059 (COFA1_HUMAN)
Collagen XVI	α1(XVI) [COL16A1]	Q07092 (COGA1_HUMAN)
Collagen XVII	α1(XVII) [COL17A1]	
Collagen XVIII	α1(XVIII) [PDB:3HON]	P39060 (COIA1_HUMAN)
Collagen XIX	α1(XIX) [COL19A1]	Q14993 (COJA1_HUMAN)
Collagen XX	α1(XX) [PDB:5KF4]	Q9P218 (COKA1_HUMAN)

Table 2.4   (*Continued*)

Collagen type	α chains [PDB code]	Chain length (Code_Speciecs)
Collagen XXI	α1(XXI) [COL21A1]	Q96P44 (COLA1_HUMAN)
Collagen XXII	α1(XXII) [COL22A1]	Q8NFW1 (COMA1_HUMAN)
Collagen XXIII	α1(XXIII) [COL23A1]	Q86Y22 (CONA1_HUMAN)
Collagen XXIV	α1(XXIV) [COL24A1]	Q17RW2 (COOA1_HUMAN)
Collagen XXV	α1(XXV) [COL25A1]	Q9BXS0 (COPA1_HUMAN)
Collagen XXVIII	α1(XXVIII) [COL28A1]	Q2UY09 (COSA1_HUMAN)

Please refer to Table 1 in Ref. [36] for prevalent polymeric molecular species of each type.

supramolecular assemblies. The supra-molecular network created through various permutation-combinations of ECM macromolecules in consequence bears functional implications (Table 2.5). Collagen α chains widely vary in size from 662 to 3152 amino acids for the human $\alpha1(X)$ and $\alpha3(VI)$ chains respectively [38].

Depending upon the structure that influences their tendency to form a typical kind of matrix, collagens may further be subgrouped as (a) fibril-forming collagens (e.g., I, II, III, V, and XI), (b) network-forming collagens (e.g., basement collagen IV and VII, VIII and X), (c) microfibrillar collagen VI, (d) fibril-associated collagen (e.g., IX, XII, XIV, XVI, XIX) with interrupted triple helix (FACIT), and (e) multiplexins (e.g., XV and XVIII), which exhibit multiple triple helix domains and interruptions. These are characterized by a central and interrupted collagen domain flanked by large non-collagenous domains at both the amino and carboxy end regions [39].

Fibrillar collagens (I, II, III, V, and XI) are the most common collagens and predominantly found in bone, tendons, ligaments and skin. They account for over 70% of the total collagen found in the body [40]. Fibril-forming collagens constitute a rather homogenous subgroup in comparison to the non-fibrillar which represents a heterogenous class of collagens. A major difference in these two subgroups is that almost the entire length (~300 nm) of the fibril forming collagen molecule is constituted of a single collagen domain, while non-fibrillar collagens contain one or more non-collagenous domains of variable sizes at both the ends and/or intermittently spread between collagenous domains. However, a member collagen

Table 2.5 Collagens and their structural and functional contribution to ECM

Collagens	Functional contribution	Physical presence
Fibril forming (I, II, III, V, XI, XXIV and XXVII)	I, II, III resist tension V, XI control fibril diameter	Tendon, ligament, intervertebral disk, bone, cartilage, blood vessels, dermis, etc.
Fibril-associated collagens with interrupted triple helices (FACIT)	IX, XII, XIV, XVI and XIX interact with other ECM components	Co-assemble with fibril forming collagens
Polygonal network forming (IV, VIII, XV and XVIII)	IV plays a role in the cell growth, migration and differentiation	Separates tissue compartments, surrounds many cell types (e.g., smooth muscle cells and nerve cells)
Membrane collagen (XIII, XVII, XXIII and XXV)	XV, XVIII:multiplexins, Unknown function	Form supramolecular assembly Associated as membrane protein
Filamentous collagen (VI)	Beaded filaments to anchor and bridge the cells with other ECM components	Ubiquitous in connective tissues VIII in cornea and vascular tissue
Short chain collagens (VIII and X)—form short-chain hexagonal network	Important in development and maintenance of tissues Unknown function	X hyaline cartilage, hypertrophic scar XIII blood vessel wall, glomeruli of kidney
Long-chain collagens (VII)	Associated with basement membrane	Adheres basement membrane with adjacent connective tissues

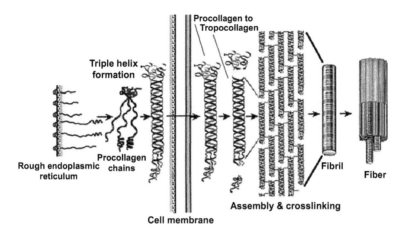

Triple helix formation

Procollagen to Tropocollagen

Rough endoplasmic reticulum

Procollagen chains

Cell membrane

Assembly & crosslinking

Fibril

Fiber

Figure 2.3 Procollagen to collagen fibers; synthesis of procollagen, release in extracellular space, conversion to tropocollagen and assembly into fibrils and fibers.

of both the classes has an intrinsic property to organize into distinct supramolecular assemblies.

Fibril-forming collagens are synthesized as procollagen molecules, which contains an amino-terminal propeptide followed by a short, non-helical, N-telopeptide, a central triple helix, a C-telopeptide and a carboxy-terminal propeptide (Fig. 2.3). Individual procollagen α -chains are then subjected to several post-translational modifications like hydroxylation of proline and lysine residues, glycosylation of lysine and hydroxylysine residues and sulfation of tyrosine residues, etc., prior to triple helix formation [41]. Both the propeptides of procollagens are cleaved during the maturation process [42]. Telopeptides contain the sites where cross-linking occurs. This process is initiated by the oxidative deamination of lysyl and hydroxylsyl residues catalyzed by the enzymes of the lysyl oxidase family [43]. HSP47 is a collagen specific molecular chaperone that is essential for the normal synthesis of collagen [44]. Fibril-forming collagen molecules fold in a C- to N-terminal direction. The correct folding of the NC1 domains at the C-terminus is crucial for collagen assembly that involves inter and intra-chain disulphide bonds.

The mechanical properties of fibril forming collagens largely depend upon covalent crosslinking. Other than disulfide bonds (existing in collagens III, IV, VI and XVI), the N^ε(γ-glutamyl) lysine isopeptide, formation of which is catalyzed by transglutaminase-2 in collagens I, III, V/XI and VII [45] mature crosslinks by lysyl oxidase; advanced glycation end products [46] and a hydroxylysine-methionine cross link that involves a sulfimine (–S1/4N–) bond has also been identified in collagen IV [47]. Lysyl-mediated cross-linking involves lysine, hydroxylysine and histidine residues. It occurs between collagen molecules from the same or different groups both at the intramolecular and intermolecular levels among types (I/II, I/III, I/V, II/III, II/IX, II/XI, and V/XI) [48–50].

Crosslinking is found to be tissue-specific rather than collagen-specific. Newly synthesized collagens form reducible, bifunctional crosslinks like aldimines and keto-imines which mature into non-reducible crosslinks. The epsilon-amino group of lysine and hydroxylysine commonly create pyridinoline and deoxypyridinoline type of crosslinks in bone and cartilage and histidinohydroxylysi-nonorleucine in skin [48, 51]. Pyrrole crosslinks in bone and arginyl ketoimine adduct called arginoline (Fig. 2.4) in cartilage are other mature crosslinks that have been identified [51]. Maturation of crosslink provides additional resistance to shear stress. Pyridinoline and deoxypyridinoline are used as urinary markers of bone resorption in patients with bone diseases such as osteoporosis [52]. Glycation increases with increasing age and that leads to the formation of several glycation end-products acting as crosslinks while compromising the elasticity in the ageing tissues. Pentosidine and glucosepane are the crosslinks respectively formed by two lysines with one arginine or one glucose. They are the most common crosslinks found in the senescent skin collagen of the elderly [53].

The triple helix is resistant to the proteolytic cleavage by pepsin, trypsin and papain. *Clostridium histolyticum* produces collagenases that cleave triple helices at numerous sites. The ability of collagens to resist cleavage by pepsin and trypsin, and their sensitivity to cleavage by bacterial collagenase, are used as research tools to identify and characterize collagens [54]. The degradation of collagen in vivo commences through MMPs, cysteine proteases (e.g.,

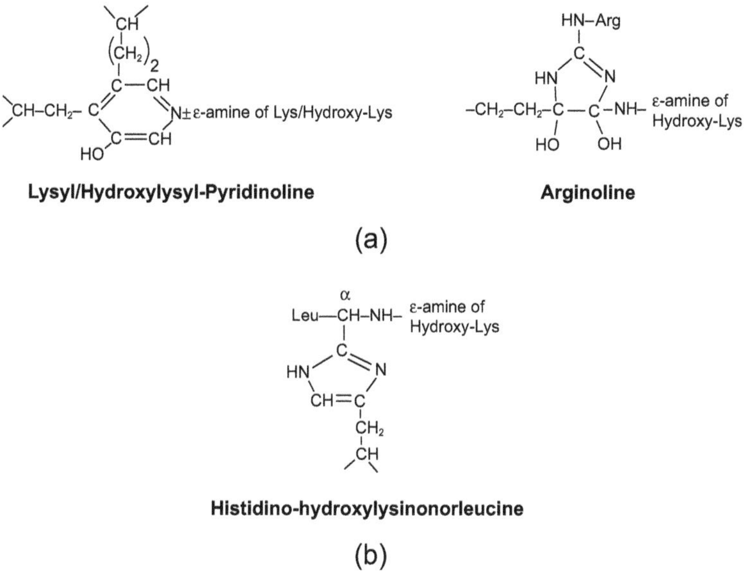

Lysyl/Hydroxylysyl-Pyridinoline

Arginoline

(a)

Histidino-hydroxylysinonorleucine

(b)

Figure 2.4 Lysyl and hydroxyl-lysyl mediated collagen crosslinks found in (a) bone and cartilage Ref. [51] and (b) skin [48, 50].

cathepsins B, K and L), and serine proteases (e.g., plasmin and plasminogen activator), etc. [55]. The helical structure of collagens imparts a spring-like property which makes them elastic. As a consequence, collagen fibrils are able to deform reversibly and can alter the mechanical properties of the ECM network [36].

The contribution of different collagens in creating a specific ECM network and its functional implications are almost established. Type III collagen for example is known to play a key role in the extensibility of tissue. It is found in abundance in embryonic tissues and also in many adult tissues like arteries, skin, and other soft and elastic organs where along with elastins they form reticular fibers [56, 57]. The prevalence of collagen III marks the initial stages of healing and scar-tissue formation. It is also an indicator of tissue maturity. As the fetal development or tissue healing proceeds, type III fibers are replaced by the stronger type I fibers [36, 58–60]. Type IV collagen on the other hand is unique in its ability to self-assemble. As a major constituent of basement membrane, it is important for the functionality and mechanical stability of basement membrane [61, 62]. However, it is not indispensible for the initiation of basement membrane assembly during early development [63]. Collagen IV molecules can self-assemble by dimerization and covalent crosslinking of the NC1 domains and by parallel and antiparallel overlapping (~30 nm) of four carboxyterminal 7S domains [64]. It is demonstrated by perturbing collagen I/V ratios within fibrils that the amount of collagen V within fibrils remains inversely proportional to their diameter [65]. This implicates collagen V in the regulation of fibrillogenesis.

Network forming collagens (IV, VII, VIII and X) create net or mesh like structures that spread in the form of a sheet as in basement membrane (type IV) or as anchoring fibrils (type VII), which link the basement membrane to the underlying ECM [66, 67]. Collagen VII is a homotrimer with the longest (450 nm) triple helix described for vertebrate collagens [68]. Fibril-associated collagens (IX, XII, XIV, XVI, XIX) do not form fibrils by themselves but assist fibrillar collagens in lateral alignments, which is crucial for supporting angular bones and joints and also for networking with other matrix molecules.

Collagens have a unique ability to resist tensile loads. The mechanical properties of fibril-forming collagens are dependent on covalent crosslinks including (1) disulfide bonds (in collagens III, IV, VI, VII, and XVI); (2) the $N1(\gamma$-glutamyl)lysine isopeptide, the formation of which is catalyzed by transglutaminase-2 in collagens I, III, V/XI, and VII [69]; (3) reducible and mature crosslinks produced via the lysyl oxidase pathway; and (4) advanced glycation end products [46]. Furthermore, a hydroxylysine-methionine crosslink involving a sulfilimine (-S1/4N-) bond has also been identified in collagen IV [70].

In general they exhibit minimal elongation (<10%) under tension which is proportionate to the straightening of the 3 dimensional arrays of packed fibers. It is not related to the elongation of individual fibers [71]. The elastic fibers on the other hand have a capability to increase their length by 150%, yet can return to their previous state [72]. Collagens also contribute significantly in attaching tendons and ligaments to the bones. At these junctions, tendons and ligaments are widened and interspaced by fibrocartilage. It represents a transformation where the aligned fibers originating from the tendon or ligament are separated by other collagen fibers arranged in a 3-dimensional network surrounding rounded cells [73]. Tendons are formed by collagen fibers aligned in parallel that strengthens them to resist unidirectional forces. It allows them to transmit forces generated by muscles to bones in an efficient manner [71]. In ligaments on the other hand, type I fibers are arranged in less parallel arrays that helps them to resist multidirectional forces. Ligaments associated with joints need to limit the motion and also provide stability to the joint [74]. It is revealed by various genetic studies that mutations in collagen genes could lead to a number of genetic disorders [75].

Small triple helical domains, i.e., collagen-like stretches, are also identified in several other proteins. They are generally secreted as soluble proteins and are therefore not included in the collagen family, which mainly encompass proteins with a much longer and predominantly multiple triple helical regions. These larger triple helical domains contribute to the overall strength, insolubility and fibril-forming tendencies of the collagen family. Emilin, Gliomedin, Ectodysplasin, Pulmonary surfactant-associated protein, Adeponectin and Ficolins are some of the well identified proteins where a collagen like domain is present [36]. Some of them where a collagen triple helix as recognition domain are called soluble defense collagens [76] whereas Gliomedin is recognized as a membrane collagen [77].

Collagens participate in cell-matrix interactions via several different receptors including integrins. They act as ligands for integrins, the cell-adhesion receptors that do not involve intrinsic kinase activities. Collagens bind to integrins via GFOGER-like sequences where O represents hydroxyl-proline [27, 78]. There are other integrin recognizing sequences in collagens, e.g., KGD in the ectodomain of collagen XVII which binds to $\alpha5\beta1$ and $\alpha v\beta1$ integrins [78]. Many other bioactive fragments achieved from proteolytic cleavage of collagens act as ligands for $\alpha v\beta3$, $\alpha v\beta5$, $\alpha3\beta1$ and $\alpha5\beta1$ integrins [79].

2.1.3.2 Elastin (elastic)

Elastin is also a structural protein that lies in the elastic fibers of connective tissue. It coils and recoils like a spring and is the major constituent of elastic fibers, the extendable element of ECM (Fig. 2.5). Elastic fibers are responsible for the property that allows tissues to endure repeated stretching and considerable deformation and still return to a relaxed state. Elastic fibers are typically large, multi-component ECM structures, spanning multiple cell diameters. Elastin paired with fibrilins and other associated glycoproteins grant elasticity and strength to the ECM through their elastic and fibrous disposition.

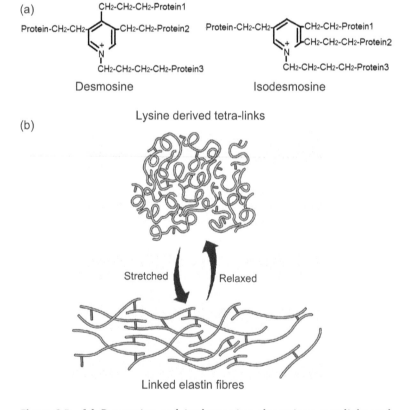

Figure 2.5 (a) Desmosine and isodesmosine, the unique tetralinks and (b) spring-like structure of elastin.

Elastin interacts with cells through microfibrils to constitute a strong but flexible matrix that helps body organs to stretch and relax to revert to their original shape (Fig. 2.5b). Such elastic fibers have a significant role in tissues like heart and lungs where rhythmic expansion is required [80]. 10–15 nm long microfibrils remain localized to the periphery of the fiber in adult tissues. Elastin is essentially found in the walls of arteries, lungs, intestines, heart and skin [81].

Elastin is an indispensable component of the vascular extra-cellular matrix that is produced during fetal and neonatal growth. Mice with decreasing amounts of elastin (from 100 to 30% normal levels) have increasingly severe vascular and pulmonary phenotypes and if it falls below 30%, they do not survive postnatal [82]. Elastin contains unique linkages called desmosine and isodesmosine formed by multiple lysines. This cross linkage is mediated by lysyl-oxidase enzyme for which copper acts as a co-factor [83]. This unusual bonding brings together the adjacent tropoelastin chains and to a great extent accounts for the elastic properties of the elastin fibers [81]. Elastin in lungs is the major component which is responsible for accommodating and balancing the air pressure variations [84]. In arteries it helps in withstanding the blood pressure by expanding during systole and contracting or reverting back to their original state during diastole [85]. In dermis the coiled elastic fibers orient preferentially at a right angle to the lines of skin tension and predominantly in the direction that allows greater stretching of the skin [86]. With age the elastic fibers lose their elasticity primarily due to reduced elastin expression and also due to conformational changes and other factors that hinder their recoiling. However, enhanced elastin and fibrillin gene expression in adult tissue may happen in response to cyclic stretching, injury, and ultraviolet radiation [87].

In contrast to the genetic diversity evident in the collagen gene family, elastin is encoded by a single gene in mammals and is secreted as a 60–70 kDa monomer called tropoelastin. Tropoelastin, the monomeric gene product, contains alternating domains of hydrophobic amino acids that contribute to the protein's elastic properties and lysine rich sequences that serves to crosslink the protein into a functional polymer (elastin). Of about 40 lysine residues in the secreted tropoelastin, all but approximately five are modified by lysyl-oxidase enzyme (LOX) to form bi-, tri-, and tetra-functional crosslinks. Such a high degree of crosslinking is responsible for the stability and insolubility of the protein. The lysines of adjacent tropoelastin chains crosslink and form desmosine and isodesmosine linkages unique to the elastin of vertebrates [88, 89]. These multivalent linkages are essential in imparting the elastic properties to elastin [81]. The elastin structure is deduced from cDNA sequencing. It shows alternating hydrophobic and crosslinking domains with the frequent presence of the VPGVG sequence.

2.1.3.3 Microfibril associated macromolecules (fibrillar)

Microfibril associated glycoproteins (MAGPs), fibulins, and elastin microfibril interface located protein (EMILIN-1) are other smaller proteins, which contribute to the creation of elastic fiber assembly (Fig. 2.6). While elastin, which makes up the bulk of the mature fiber, arises from a single gene, the composition and genesis of microfibrils that serve to link cells to elastic fibers in the extracellular matrix is more complex. Microfibrils are most eminent for their association with elastic fibers but they can also be found without elastin in the ciliary zonules of the eye and in the circulatory system of

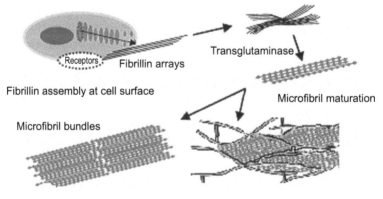

Fibrillin assembly at cell surface

Fibrillin arrays

Receptors

Transglutaminase

Microfibril maturation

Microfibril bundles

Inter microfibril crosslinks

Tropoelastin deposit to create elastic fiber

Figure 2.6 Elastic fiber formation (partly adapted with permission from Kielty et al., *J Cell Sci*, 2002, **115**, 2817–2828).

invertebrates where they have a mechanical role to play [90]. With quick freeze deep-etch (QFDE) microscopy, microfibrils appear as tightly packed rows of bead-like subunits showing a periodicity of 10–14 nm along the microfibril length. When microfibrils are in bundles a larger bead is also recognized at regular 50 nm intervals, which tends to be aligned with those from adjacent microfibrils [91]. The coordinated expression pattern of fibronectin, fibrillin-1, MAGP-1 and fibulin-5 except fibrillin-2 and MAGP-1, implies that each protein is necessary for normal elastic fiber assembly. The proteins are temporally arranged into fibers in that order: fibronectin, fibrillin-1, MAGP-1, fibulin-5, and elastin in a variety of elastin producing cells.

(1) *Microfibril associated glycoproteins (MAGPs)*: MAGPs are small glycoproteins (~20 kDa for the mature secreted form) with no repeating motifs [92]. The two members of the MAGP gene family (MAGP-1 and -2) are related through a 60-amino acid sequence in the middle of the molecule that shares a precise alignment of 7 cysteine residues [93]. Sequence comparison shows that MAGP-1 is a highly conserved protein. It undergoes several posttranslational modifications that may influence its associations with other microfibrillar components. The O-linked glycosylation and tyrosine sulfation at the amino-terminal region prepare this site as a major amine acceptor for transglutaminase mediated self-assembly [94]. Detailed study in knockout models suggests that fibrillin-2 and MAGP-1 are the predominant players in microfibril biology during embryonic and fetal aortic development, whereas fibrillin-1 and both MAGP-1 and –2 are the major components of vascular microfibrils postnatally and into adulthood. While both MAGP-1 and MAGP-2 have been localized to elastin-associated as well as elastin-free microfibrils, MAGP-2 exhibits a pattern of tissue localization and developmental expression that is more restricted than that of MAGP-1 [95, 96]. MAGP-1, however, is not necessary for normal elastic fiber assembly but important for other processes of tissue homeostasis or differentiation. It has been shown to bind members of the TGF-β growth factor family [97].

(2) *Fibrillin*: Microfibrils consist mainly of fibrillin, a 350-kDa glycoprotein. The primary structure of fibrillin is dominated by calcium binding epidermal growth factor (EGF)-like domains that develop a rod-like structure in the presence of calcium [98]. There are three fibrillins identified in human genome but fibrillin-3 is missing in the mouse genome possibly due to chromosome rearrangement [99]. Fibrillins are highly homologous modular structures with repeating calcium-binding epidermal growth factor (EGF)-like domains interspersed between 8-cysteine domains similar to those found in the latent transforming growth factor-beta-binding protein (LTBP) family [100]. Aggregation of Fibrillin molecules via disulphide bonds is postulated to be an early step in microfibril assembly. The N-terminal region of each protein directs the formation of homodimers and that disulphide bonds stabilize this interaction. Pulse-chase analysis demonstrated that dimer formation occurred intracellularly, suggesting that the process of Fibrillin aggregation is initiated early after biosynthesis of the molecule [101]. Fibrillins interact and bind with tropoelastin via the lysine side chain. It shows no binding with mature elastin whose lysine side chains have been exhausted to form crosslinks. It has further been observed that fibrillin constructs do not interact with tropoelastin in the solution phase, suggesting that binding of tropoelastin to a solid substrate exposes a cryptic binding site [102]. Besides acting as a connecting bridge between cells and elastic fibers, another important function of fibrillin containing microfibrils is to bind and sequester growth factors into the ECM [103]. All three fibrillins contain the integrin-binding RGD sequence and fibrillin-1 and fibrillin-2 have been shown to interact with $\alpha v\beta 3$ and $\alpha 5\beta 1$ integrins as well as with heparin sulfate proteoglycan on the cell-surface [104]. Thus, fibrillins have the potential of providing instructive signals to cells either indirectly through binding with growth factors or directly by interacting with signaling receptors on the cell surface. The importance of fibrillins has been demonstrated through knockout mice. It is observed that mice lacking fibrillin-1 gene (fbn1 -/-) die within two weeks of birth from vascular and

pulmonary complications, including ruptured aortic aneurysms, impaired breathing and diaphragmatic collapse. The elastic fibers in the aorta are abnormally thin and fragmented in these mice [105]. Mice deficient in both fibrillins (fbn1 -/-; fbn2 -/-) die in utero. However, mice lacking the fibrillin-2 gene (fbn2 -/-) develop syndactyly, but show no defects in the vascular and pulmonary systems and have a normal lifespan.

(3) *Fibulins*: Fibulins, another microfibrill-associated protein are a family of seven that contain a calcium binding epithelial growth factor (EGF)-like domains [106–108]. They are often found in association with elastin fibers and are also known to bind to multiple components of the ECM and basement membrane. Fibulins are 50–200 kDa in size [109] with a distinctive COOH-terminal domain. All fibulins except fibulin-6 and fibulin-7 are found in elastic tissues, with fibulin-2 and -4 at the interface between the central elastin core and its surrounding microfibrils, fibulin-1 located within the elastin core, and fibulin-5 associated with microfibrils [110].

(4) *Elastin microfibril interface located proteins (EMILIN)*: Elastin microfibril interface located proteins are found in elastin-rich tissues like skin and arteries and is localized to the interface between amorphous elastin and microfibrils [111]. EMILIN-1 from this newly defined EMILIN/multimerin family was previously known as gp115. EMILIN-1 expression is highest at the early stages of aortic development (E14–18) then drops to low levels at birth and during the postnatal period until elastin and collagen production begins to decline at approximately P21-P30. At this point, EMILIN-1 levels increase and persist at high levels into the adult period [20]. EMILIN-1 binds elastin in solid-phase binding assays and both elastin and fibulin-5 in immunoprecipitation assay, suggesting its deposition takes place only after the fibrillar scaffold is in place [112]. Absence of EMILIN-1 increases TGFb signaling and the vascular phenotype can be reversed by inactivating one TGF-b allele (emilin- -/- ; tgf-b +/- mice) [113].

Elastic fiber assembly is a complicated process involving a co-ordinated expression of multiple different proteins and enzyme activities. As mentioned before, elastic fibers consist of two morphologically distinct components: elastin and microfibrils. During development, the microfibrils form bundles that appear to act as a scaffold for the deposition, orientation, and assembly of tropoelastin monomers into an insoluble elastic fiber. Three stages are identified during elastic fiber formation. First, microfibrils are assembled. Then, immature elastic fibers appear, which consist of small deposits of amorphous material, distributed among the bundles of microfibrils (Fig. 2.6). Slowly, the fibers acquire a more filamentous final quality with an increase in size of the fibers and a parallel disappearance of deposited material [114]. The initial formation of small aggregates is termed 'microassembly', while the transfer of the older, sometimes larger aggregates to existing fibers is termed 'macroassembly' [115]. Pulse-chase immunolabeling of the fibroblast-like rat fetal lung fibroblast (RFL-6) cells demonstrates that tropoelastin globules aggregate in a hierarchical manner, creating progressively larger fibrillar structures. Although microfibrils can assemble independent of elastin, tropoelastin monomers do not assemble without the presence of microfibrils. Live imaging studies in immortalized ciliary body pigmented epithelial (PE) cells [116] have highlighted the active role of the cell in elastic fiber assembly for coordinating and directing the proteins to appropriate locations.

Humans are exquisitely sensitive to reduced elastin expression, developing profound arterial thickening and markedly increased risk of obstructive vascular disease [117]. In adult tissue, elastic fibers are difficult to repair and elastin often does not polymerize or form functional three-dimensional (3D) fibers [118, 119]. It is observed that tropoelastin pre-mRNA is transcribed at the same rate in neonates and adults, but fully processed transcripts are unstable and seem to limit the protein production in mature tissue [120]. Since the replenishment of elastin slows down with age, the body tissue starts losing their elasticity. Other than age, the elastin may be damaged by factors like smoking, ultraviolet radiation, stress, hormone or injury.

2.1.3.4 Laminin (adhesive)

Laminin represents cell adhesion molecules that comprise a family of heterotrimeric glycoproteins found predominantly in basement membranes (BM) [1, 2]. The unique cross shape of laminin makes it an efficient interconnecting molecule (Fig. 2.7). As an essential component of basement membrane Laminin is present throughout the body, creating compartments and surrounding the muscles, Schwann and fat cells [121]. It also contributes in the adhesive, migratory and signaling functions of ECM [122, 123]. Laminin 1 has the propensity to self-associate into polymers in the presence of calcium [124]. Like collagen IV, it can self-assemble into mesh-like

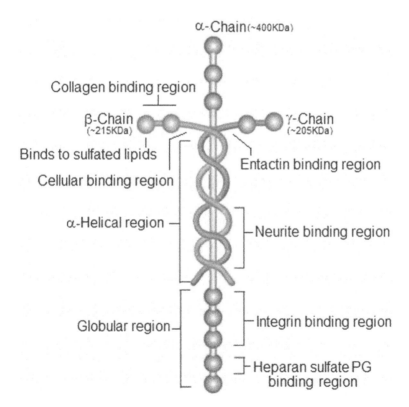

Figure 2.7 Laminin molecule.

polymers. It can initiate BM assembly at peri-implantation stages even in the absence of collagen IV [63, 125]. The latter, however, is required for conferring barrier functions and maintaining BM integrity associated with increasing mechanical demand as the embryo grows.

Laminin exists as a branched molecule that consists of three genetically distinct polypeptides, the alpha, beta and gamma chains. Laminins are heterotrimeric glycoproteins consisting of α, β and γ chains, respectively encoded by the genes *LAMA1-5*, *LAMB1-3* and *LAMC1-3* [126]. In vertebrates, five α, three β, and three γ chains have been identified that represent distinct gene products (Table 2.6).

Laminin also forms a stable high affinity equimolecular complex with the carboxy terminal domain of nidogen through epithelial growth factor (EGF) like motif at its gamma chain [127]. Nidogen apparently is an essential component for bridging the laminin

Table 2.6   Old and new nomenclature of laminin

Old nomenclature	Chain composition	New nomenclature
Laminin-1	$\alpha1\beta1\gamma1$	Laminin-111
Laminin-2	$\alpha2\beta1\gamma1$	Laminin-211
Laminin-3	$\alpha1\beta2\gamma1$	Laminin-121
Laminin-4	$\alpha2\beta2\gamma1$	Laminin-221
Laminin-5 / Laminin-5A	$\alpha3A\beta3\gamma2$	Laminin-332 / Laminin-3A32
Laminin-5B	$\alpha3B\beta3\gamma2$	Laminin-3B32
Laminin-6 / Laminin-6A	$\alpha3A\beta1\gamma1$	Laminin-311 / Laminin-3A11
Laminin-7 / Laminin-7A	$\alpha3A\beta2\gamma1$	Laminin-321 / Laminin-3A21
Laminin-8	$\alpha4\beta1\gamma1$	Laminin-411
Laminin-9	$\alpha4\beta2\gamma1$	Laminin-421
Laminin-10	$\alpha5\beta1\gamma1$	Laminin-511
Laminin-11	$\alpha5\beta2\gamma1$	Laminin-521
Laminin-12	$\alpha2\beta1\gamma3$	Laminin-213
Laminin-14	$\alpha4\beta2\gamma3$	Laminin-423
and	$\alpha5\beta2\gamma2$	Laminin-522
Laminin-15	$\alpha5\beta2\gamma3$	Laminin-523

and collagen IV network in the basement membrane. Mutations in laminin are implicated in human diseases like Pierson syndrome, Merosin congenital muscular dystrophy and Junctional epidermolysis bullosa [31, 128]. Many laminins self-assemble to form networks that remain in close association with cells through interactions with cell surface receptors. It provides structural support, acts as a selective barrier, and modulates signaling cues for adjacent cells. It is first synthesized by the primitive endoderm and trophectoderm, in the peri-implantation mouse embryo at the blastocyst stage.

Although, laminin subunits can assemble in 45 different possible ways only 15 types that exist in cruciform, rod-, T- or Y-shaped have so far been identified [129–131] in mammals. All laminin chains share a common domain structure with a number of globular and rod-like domains. The trimers are named according to their chain composition of α, β and γ chains, e.g., laminin composed of $\alpha3\beta3\gamma2$, formerly known as laminin-5, is now called laminin-332. Similarly, laminins that are composed of $\alpha1$, $\beta1$ and $\gamma1$ chains and $\alpha4$, $\beta2$, and $\gamma3$ chains are called LM-111 and LM-423 respectively [132]. As some chains are components of multiple laminins, e.g., laminin $\alpha5$ chain is a component of LM-511, LM-521 and LM-523, characterizing the role of individual laminins is difficult as mutations may affect multiple laminins. Laminin-111 ($\alpha1\beta1\gamma1$) and laminin-511 ($\alpha5\beta1\gamma1$) are the earliest isoforms found in embryonic BM, and absence of both leads to early lethality of mouse embryos with defects in primitive endoderm differentiation and epiblast polarization [133]. Blood vessels are found to have $\alpha4$, $\alpha5$, $\beta1$ and $\gamma1$ chains of laminin, suggesting the presence of laminin-8 and 10, which is further confirmed by demonstrating that vascular endothelial cells produce these isoforms [122, 134, 135].

2.1.3.5 Fibronectin (adhesive)

Fibronectin (FN) is another extensively studied adhesive glycoprotein of ECM [136]. It is the key component of ECM that facilitates connections not only among ECM macromolecules but also links ECM to cells. Having binding sites for both collagen and GAGs, it can crosslink these components in the matrix. Depending on the cell source the carbohydrate content in FN is found to be 4–9% [137]. Gene knockout studies have shown that fibronectin is essential for the organization of heart and blood vessels. In the absence of fibronectin, aortic endothelial cells do not organize into tubes and, as a consequence, blood vessels do not form [138].

Fibronectin is a multi-domain V-shaped protein that consists of two nearly identical ~250 kDa subunits joined together by a pair of disulfide bonds near their carboxyl termini. Each subunit is folded into a series of functionally distinct rod like domains separated by regions of flexible polypeptide chain (Fig. 2.8). Sequence analysis shows presence of three different repeats termed as FN-I, FN-II, and FN-III. Each subunit contains 12 FN-I, 2 FN-II, and 15 to 17 FN-III repeats. Type I repeats are about 40 amino-acid residues in

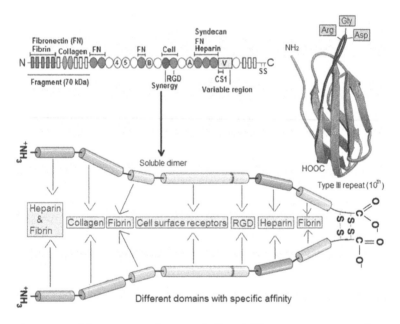

Figure 2.8 Fibronectin with types I, II and III repeats represented as rectangle, oval and circle and its soluble dimer exhibiting different domains with specific binding affinities. The NMR (nuclear magnetic resonance)-based three-dimensional structure of type III fibronectin repeat, which is found in many other proteins and carries the RGD peptide.

length and contain two disulfide bonds; type II repeats comprise a stretch of approximately 60 amino acids and two intra-chain disulfide bonds. Type III repeats are about 90 residues long, with maximum glycosylated sites and no disulfide bond [2]. While type I or type II domains contain cross-linked beta-strands, the type III domain has seven beta strands existing in sandwich format. Fibronectin possesses one RGD sequence located in the tenth type III repeat, which is recognized by many members of the integrin family [139]. Besides the RGD peptide on FN-III-10, FN-III-9 also contains a short peptide that binds to integrin $\alpha 5$-$\beta 1$. Also called synergy site, it is located about 32 Å from the RGD peptide and acts together with the FN-III-10 binding site to enhance integrin-mediated cell adhesion. Though fibronectin protein is a single gene product, yet alternative splicing of its pre-mRNA may create as many as 20 variants in human FN [140]. Cellular FN consists of a much larger and more heterogeneous group of FN isoforms that result from cell-type specific and species-specific splicing patterns. Based on structural studies it has been proposed that the unfolding of the FN-III domain provides the elasticity to FN fibrils. It is suggested that unfolded and stretched FN-III domains expose buried binding sites that serve as focal points for the assembly of FN into fibrils. These buried binding sites, called cryptic sites for fibrillogenesis have been proposed to exist in FN-III-1, FN-III-2, FN-III-7, FN-III-9, FN-III-10, and FN-III-13-15 domains. Other than fibrillogenesis these cryptic sites may also be involved in the interaction of FN with other ECM proteins.

In plasma, FN exists as a soluble dimer, but in the ECM it is found as an insoluble multimer. Soluble FN (plasma FN or pFN) produced by hepatocytes is in abundance (300 mg/ml) and circulates in plasma in its inactive form [1]. Many adherent cells like fibroblasts, chondrocytes, endothelial cells, macrophages, as well as certain epithelial cells synthesize insoluble fibronectin (cellular FN or cFN), which gets assembled into an extracellular fibrillar network stabilized by disulfide bridges [136]. However, in contrast to the classical self-assembly of collagen, FN polymerization is cell driven where several regions of the molecule are involved in the assembly process. Electron microscopic analyses of natural thin fibrils (5–18 nm diameter), made by fibroblasts in culture, clearly indicate an ordered arrangement and suggest a model in which extended protomers (130 nm long) are arranged end-to-end with an overlap of about 14 nm [141]. Fibronectin fibrils exhibit elastic properties and cells can presumably stretch FN fibrils up to four-fold in their relaxed length. The creation and deposition of insoluble FN fibrils into the ECM is a tightly regulated, cell mediated process termed FN fibrillogenesis or FN matrix assembly [142–144]. Dimer state and integrin interaction of FN-RGD seems to be important for fibril formation as both the monomeric FN [145] and the recombinant fibronectin lacking RGD motif [146] fail to form fibrils. It has been further demonstrated, using hepatocytes, that $\alpha5$-$\beta1$ integrin recognizes FN in its unfolded state, which seems to be crucial for binding and therefore fibrillogenesis [147]. Once assembled, the FN matrix is expected to influence tissue organization by contributing to the network formation of other ECM proteins [148].

FN binds to a number of biologically important molecules that include heparin, collagen/gelatin, and fibrin. Several distinct structural and functional domains mediate these interactions. There are two major heparin-binding domains in FN that interact with heparin sulfate proteoglycans, the strongest being in the C-terminal part (Heparin II) and a weaker binding domain is located at the N-terminal end of the protein (Heparin I). FN can also be a ligand for many members of the integrin family that include $\alpha5\beta1$, $\alpha v\beta1$, $\alpha v\beta3$, $\alpha v\beta5$, $\alpha v\beta6$, $\alpha II\beta3$, and $\alpha8\beta1$ integrins [139, 149]. $\alpha5\beta1$ integrin is expressed by many cell types and is probably the major fibronectin receptor in several of these. It was the first integrin identified to be involved in fibronectin network formation. Antibodies to $\alpha5$ or $\beta1$ can inhibit the polymerization of fibronectin in fibroblast cultures [150]. Mere binding of fibronectin to the cell surface by any integrin is not sufficient. It is observed that a connection of the β subunit of integrin to actin filament is also required for fibronectin fibril formation on the cell surface [151]. As a result, most cell types in the body can adhere to FN via integrin receptors and by this means FN gets involved in several different biological processes like cell adhesion, cell migration, tissue repair, embryogenesis, blood clotting, wound healing, etc. Thus, fibronectin is able to interact with cytoskeleton through integrins; ECM through collagen, tenacin, fibulin and thrombospondin; with circulating coagulation factors like fibrin through transglutaminase or factor XIII; with itself through cryptic binding sites and with viruses through the Heparin II domain.

2.1.3.6 Matricellular (anti-adhesive)

Matricellular represents a group of extracellular proteins, which can modulate cell function but are not crucial for the structural integrity of ECM [152, 153]. Their ability to bind cell-surface receptors and also to other ECM biopolymers like collagens, hormones, enzymes, growth factors, etc., equip them to control the cellular response by managing the availability of cell-effectors. The characteristics which distinguish the matricellular proteins are (i) high levels of expression during development and in response to injury; (ii) induction of de-adhesion or counter-adhesion in contrast to adhesive properties exhibited by most of the matrix proteins [154, 155]; (iii) binding to many cell-surface receptors and components of the extracellular matrix, growth factors, cytokines, and proteases and (iv) apparently normal or subtle phenotype difference that is observed in mice with a targeted disruption (knockout) of some matricellular protein genes [153].

Matricellular proteins like thrombospondin (TSP), osteonectin, also known as 'Secreted Protein, Acidic and Rich in Cysteine' (SPARC) [156], tenascin [157], osteopontin [158], etc., are also called "anti-adhesive proteins" because they can induce rounding and partial detachment of some cells in vitro [152, 159]. TSPs particularly showed unusual functional diversity by displaying TSP-1 as angiogenic and TSP-2 as an anti-angiogenic molecule [160, 161]. The role of TSPs in wound healing, ischemia, cardiac remodeling, foreign body response and also in bone regeneration are also reported [162]. CCN1 (formerly named CYR61 cysteine-rich angiogenic inducer 61; NM_001554) is another secreted, cysteine-rich matricellular protein. It is also known to regulate cellular response by binding to integrin receptors and heparan sulfate proteoglycans [160] and its absence in *Ccn1*-null mice impairs the cardiac valvuloseptal morphogenesis, resulting in severe

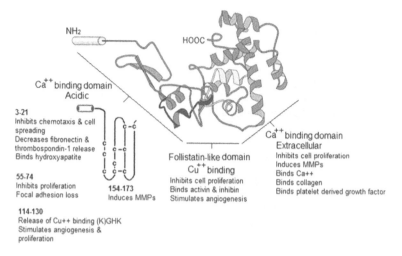

Figure 2.9 Structure of human SPARC (Secreted Protein, Acidic and Rich in Cysteine) with major domains defined by residue numbers. Their location, secondary structure, and function on cultured cells are also notified. (C: cysteine, MMPs: matrix metalloproteinase, Ca^{++} calcium, Cu^{++} copper, Adopted and combined from *Nat Med*, 1997, **3**, 144–146, and Figure 1 of *Nat Rev Endocrinol*, 2010, **6**, 225–235.)

atrioventricular septal defects (AVSD) [163]. The human SPARC consists of 286 residues divided into three distinct domains (Fig. 2.9). The N-terminal domain (residues 1–52 after a 17-amino-acid signal sequence) is an acidic region rich in Asp and Glu. Domain I binds several calcium ions with low affinity [164] and interacts with hydroxyapatite [165]. It has therefore been implicated in the mineralization of cartilage and bone. This N-terminal domain contains the major immunological epitopes of SPARC [166, 167]. SPARC is also a predominant glutaminyl substrate (amine acceptor) for transglutaminase in the chondrocyte matrix and is co-expressed with transglutaminase in maturing cartilage [168]. The formation of SPARC oligomers or complexes by transglutaminase has been identified in cartilage and is believed to stabilize the extracellular matrix of this tissue [169]. SPARC is widespread in connective tissue and is also found in the basement membrane [170, 171] where it presumably binds to collagen IV.

Expression of matricellular proteins is high during embryoge-nesis, but almost absent during normal postnatal life. However, at the instance of injury their expression is resumed. Several members of the family including tenascin-C, osteopontin, and osteonectin are found up-regulated after cardiac injury [172]. Matricellular proteins are known to induce de-adhesion by rapid transition to an intermediate state of adhesiveness. This transient state is characterized by loss of actin containing stress fibers and restructuring of the focal adhesion plaque that includes loss of vinculin and α-actinin, but not of talin or integrin. TSP1 and 2, tenascin-C and SPARC, each of these metricellular proteins employs different receptors and signaling pathways to achieve this common morphologic endpoint [173]. The process of cellular de-adhesion is potentially important for the ability of a cell to participate in morphogenesis and to respond to damaging stimuli. Furthermore, it has been suggested that the strong adhesion prevents the cell from releasing its cytoskeleton-ECM linkages, whereas weak adhesion does not generate the contractile force necessary for directed cell movement. It is the intermediate state of adhesion that favors cell motility [174], which is very much needed in areas of remodeling required during embryogenesis, wound healing, and inflammation. Increased expression of matricellular proteins during development and in response to injury, suggests that the promotion of the intermediate adhesive state for facilitating cell migration and ECM remodeling may be one of the important functions of these molecules [173]. The active site of each of these matricellular proteins has been localized. In SPARC the focal adhesion reorganization is stimulated by two sequences located at different domains [175]. In TSP1 this activity lies in the NH2-terminal heparin-binding domain (HBD) [176].

Matricellular proteins play an important role in mediating and presenting new information to the cells. Having multiple binding domains for different types of ECM molecules and enzymes they are capable of tethering/sequestering growth factors (e.g., VEGF or PDGF with SPARC), binding ions (e.g., Ca^{2+} with OPN), inhibiting proteases by direct binding (e.g., serine proteases or MMP3 with TSP1), clearance of proteases by endocytosis (e.g., MMP2, LRP with TSP2), and activating cytokines (e.g., latent TGF-β1 with TSP1). It is important to note that some matricellular proteins, such as TSP1 and TSP2, bind to almost all types of integrin receptors whereas no receptor has yet been identified for SPARC or osteonectin [172]. Thus matricellular proteins seem to assist in keeping the balance/homeostasis in the dynamic relation that exists between cells and its ECM.

2.1.3.7 Matrikines and matricryptins

Matrikines and matricryptins are the peptides and proteins which are either released or exposed as the new binding sites through controlled unfolding or partial proteolysis of ECM molecules. The smaller peptide fragments or the new unmasked sites, also called cryptic sites, allow fresh interactions which may alter cellular behavior. Released fractions (matrikines) or unfolded subdomains (matricryptins) of various ECM biopolymers, capable of signaling the cells through integrins or other growth factor receptors belong to this class [177]. Offered as new ligands on ECM for interaction and binding, under specific physiological conditions, they generally involve different mechanisms compared to the originating molecules for engaging cells and/or partner ECM polymers. These interactions can modulate proliferation, migration, protease production or even apoptosis [178]. For example, EGF-like repeats present in tenascin-C and laminin bind to EGF receptors and enhance cell motility [179, 180]. In laminin 332, this activity is observed only after cleavage by matrix metalloproteinase (MMPs) [181]. Unlike the entire native molecule, the released matrikines possess low binding affinity to their receptors and are often presented in multiple valence states that likely increase avidity to receptors leading to distinct consequences [182]. As most of the matricryptins derived from collagens, proteoglycans and glycosaminoglycans exhibit anti-angiogenic and anti-tumor properties, they have a potential to be used as disease markers or even as drugs [183].

Matrkines may also be referred to as proteolytically processed intermediates or fragments of ECM macromolecules that interact and co-ordinate with other ECM companions to ensure a homeo-static state during development, tissue repair and wound healing [184].

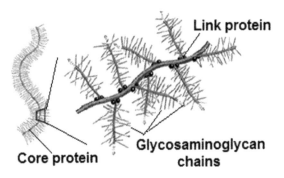

Figure 2.10   Typical proteoglycan structure.

2.1.3.8 Proteoglycans

Proteoglycans (PGs), earlier called mucoproteins, are other major contributor to the morphological state of the extracellular matrix (ECM). They represent the most diverse group of multifunctional, structurally unique heteropolymers [189]; PGs correspond to a class of genetically unrelated, multidomain proteins carrying multiple glycosaminoglycan (GAG) side chains (except hyaluronic acid) [185–187]. These glycosaminoglycan (GAG) chains are covalently attached to a specific protein core and may vary widely in their size and number. Morphologically they look like a bottle brush (Fig. 2.10). The type and number of units of GAG largely determine the properties of the PG [188].

PGs may differ in their core protein and also in the composition of associated GAGs, making them highly functional macromolecules of ECM. Type and arrangement of GAGs over the core protein apparently helps the protein to sense its environment. It is likely that PGs act as sensors in the complex matrix of ECM. The GAGs owing to their hydrophilicity with the designated number of polar groups are capable of responding conformationally. Their conformation is expected to change with the surrounding polarity and hydrophilicity and this alteration could get propagated to core protein. Such molecular transmittance would depend on the intensity and frequency of accruing changes. Accordingly it will alter and modulate first the side chain conformation and then the back bone of the core protein unfolding different cryptic sites. Dominance of PGs in glomerulus and in the basement membrane of almost all the vital tissues substantiates this hypothesis.

PGs along with other GAGs, collagens and matrikines fill the interstitial spaces between the cells. They also form part of the basement membrane and often display a functional association with the cell surface receptors [188, 189]. Substantial evidence exists to show that PGs act as a glue between various collagen and glycoprotein networks through specific interactions with single components. The protein core of decorin for example, interacts with collagen VI [190], Fibronectin [191] and thrombospondin [192] while biglycan interacts with collagen via the glycosaminoglycan side chains [193].

PGs are responsible for regulating and stabilizing the collagen network and the hydration state of ECM. Their hydration capacity and specific distribution in the collagen matrix imparts the tissue with an ability to resist compressive forces. This property is best reflected by the PGs of articular cartilage [188]. Formerly considered as mere fillers of ECM, PGs are now revealed to be an active player in keeping the homeostatic balance of ECM-cell dynamics. PGs are present not only in the extracellular and intracellular space but also span through cellular membranes. This makes them ubiquitously present biopolymers and implicates them in the complete spectrum of inducer to effecter processes. Besides mediating various cell-ECM and cell-cell interactions, appropriately arranged GAGs in PGs are crucial for tissue plasticity. They also assist in maintaining the viscoelasticity of blood vessels and the tensile strength of the skin and tendons.

More than 30 different proteoglycans, most of which reside in the ECM, are reported so far [194, 195]. Earlier PGs were classified according to the major GAG chains they carry. GAGs are linear, sulfated, negatively charged polysaccharides, which can be divided into two classes: (i) sulfated GAGs; comprising chondroitin sulfate (CS), dermatan sulfate (DS), keratan sulfate (KS), heparan and heparin sulfate (HS) and (ii) non-sulfated GAGs such as hyaluronan or hyluronic acid (HA) [196]. However, based on typical features of the core protein, a simplified classification into two subfamilies of glycoproteins and proteoglycans is also proposed [185] by Aumailley and Gayraud (Table 2.7) [197].

Table 2.7 Some well identified glycoproteins and proteoglycans of ECM classified on the basis of GAGs by Aumailley and Gayraud [197]

Glycoproteins	Proteoglycans
Interstitial connective tissue	*Small leucine-rich proteoglycans*
Fibronectins	Decorin
Tenascins	Biglycan
Fibrillins	Fibromodulin
Elastin	Lumican
Microfibril-associated	Epiphycan
Matrilins	*Modular proteoglycans: non-hyaluronan*
Thrombospondins	*binding* Perlecan
Basement membranes	Agrin
Laminins	Testican
Nidogen/entactin	*Modular proteoglycans: hyalectans*
Fibulin	(hyaluronan- and lectin-binding)
	Aggrecan
	Versican
	Neurocan
	Brevican

PGs also contribute to the structural integrity in tissues like cartilage and cornea. The complexity in PGs is mainly caused by the size and number of GAG chains and also due to the variation in the core protein structure and characteristics. For simplification, we prefer to classify PGs into three categories: (i) small leucine-rich proteoglycans (SLRP); (ii) modular proteoglycans and (iii) cell-surface proteoglycans.

(i) *Small leucine-rich proteoglycans (SLRP)*: A protein core composed of leucine-rich repeats (LRRs) flanked by two conserved cysteine-rich regions is the main characteristic of SLRPs. Represented by 17 genes, SLRPs can be further classified into five distinct families on the basis of conservation and homology of protein and their genes and also the presence of characteristic N-terminal Cys-rich clusters with defined spacing and chromosomal organization [195]. In addition to being structural proteins, SLRPs network with other ECM allies for signal regulation. Decorin, biglycan, fibromodulin, epiphycan and lumican are some of the commonly found SLRPs that exist associated with collagen fibrils. They have a small protein core (36–42 kDa) of leucine-rich repeats flanked by cysteine clusters and a carboxy terminus domain. Depending on the member, one to four glycosaminoglycan chains remain attached to its amino-terminal domain. The protein core of decorin, lumican or fibromodulin is compact and horseshoe-shaped that helps in protein-protein interaction [198]. The crystal structure of the protein core of decorin has confirmed the presence of conserved N-terminal and C-terminal cysteine rich regions which cap the similar internal repeat structures [199]. The non-covalent and presumably reversible interactions of core protein of SLRP with other ECM proteins are reported to be important for functional network formation. Several SLRPs interact with fibrillar collagens including types I, II, III, V, VI, and XIV [200]. SLRPs, along with fibronectin and other ECM partners, contribute to collagen fibrillogenesis, a multi-step process involving intermediates undergoing linear or lateral growth to become mature fibrils [201]. Lumican, for example prevents lateral fibrillar growth in the cornea while mediating homogeneous distribution of small-diameter fibrils to favor transparency [202].

(ii) *Modular proteoglycans*: The modular PGs represent a highly gly-cosylated, heterogeneous group characterized by an elongated protein chain that possesses different functional domains. This group is further divided into two subfamilies: non-hyaluroan binding and the hyaluronan and lectin-binding PGs (also called hyalectans) [203].

   (a) *Non-hyaluronan   binding*:   Two   members   of   non-hyaluronan-binding modular proteoglycans, the perlecan and agrin are specific for basement membranes. The core protein of chondroitin sulfate proteoglycan Bamacan from basement membrane is a five domain structure [204]. The non-HA binding modular PGs may exist as such or produced by proteolytically processed hyalectans where HA binding domains and/or GAG binding domains are disintegrated, e.g., Neurocan and brevican, the two CNS hyalectans are found to exist as full length PGs or in its fragmented form which lacks the HA binding N-terminal domain.

(b) *Hyalectans*: The hyalectan subgroup like versican, aggrecan, neurocan and brevican possess a large protein core (100–370 kDa). The N-terminal domain of hyalectans binds with hyaluronic acid (HA) and the C-terminus, which is rich in epidermal growth factor (EGF)-like repeats, interacts with the lectins [185]. Most of the GAG chains in hyalectan-PGs are confined to the central protein domain, the size of which dictates the associated number of GAG chains that can vary from 3 in brevican to 100 in aggrecan. Versican is the largest member of the hylectans gene family and is capable of interacting with a number of ECM components like HA, Col-1, fibrillin-1, fibronectin, P and L-selectin, CD44, TGFR, integrin b1, TLR2 [205]. Versican and CD44, the two members of hyalectans, are widely distributed whereas aggrecan is expressed mostly in cartilage and neurocan, also known as brevican, are brain tissue specific [197].

(iii) *Cell-surface/transmembrane proteoglycans*: The cell surface PGs [206] are also of two types: membrane spanning, e.g., syndecans and betaglycans; and glycosylphosphatidyl-inositol (GPI) anchored PGs, e.g., glypicans and spliced form of brevican.

(a) *Syndecans*: These PGs being membrane spanning are often described as co-receptors due to their presence alongside other receptors, such as integrins and high affinity tyrosine kinase growth factor receptors [207]. Most cell types, with the exception of erythrocytes, express at least one out of the four members of the syndecan family. The core protein of syndecan ranges from 20–40 kDa with the heparin sulfate (HS) as the principal GAG. However, syndican-1 and 3 may also possess chondroitin sulfate (CS). Syndecans can interact with actin-associated proteins and signaling molecules like protein kinases [208]. Functional actin cytoskeleton is a requirement for extracellular matrix (ECM) assembly and all syndecans have been shown to interact with actin-associated proteins. This confirms the syndecans' role in ECM remodeling [209].

(b) *Glypicans*: Unlike syndecans, which span all through the membrane and act as receptor/co-receptor, glypicans are linked to the cell membrane through phosphatidylinositol linkage. In mammals, glypicans represent a family of six members each characterized by HS-GAGs linked to a core protein of 60–70 kDa with 14 highly conserved cysteine residues. In adults, glypican-1 is ubiquitously expressed, whereas glypican-6 is found mainly in the heart, kidney, liver, ovaries, and intestine. By contrast, glypican-3, -4, and -5 are located exclusively in the CNS. Glypican-2 is present in the CNS only during embryonic development. Glypicans are involved in the regulation of various signaling pathways including that of Wnt, fibroblast growth factor, Hedgehog, bone morphogenic protein, Slit, and insulin-like growth factor signaling [210]. They are part of the pericellular matrix and mostly remain confined to the pericellular region. Thus they are the prime mediators that connect cells with the ECM.

While existing in a secreted form or cell membrane bound or ECM bound state, PGs can mediate a variety of functions like ECM assembly and remodeling, cell adhesion and motility. Heparin sulfate proteoglycans (HSPGs) which may be secreted or remain ECM bound (like perlecan) or membrane-tethered (like glypican and syndican) are discovered to be growth factor modulators [211]. It is further observed that variations in the heparin sulfate side chain of HSPGs confer affinities to different fibroblast growth factors (FGF), e.g., presence of L-iduronic acid favors binding to FGF2 but not FGF10 [212].

Though PGs like perlecan, decorin and versican are widely expressed in many different tissues, some PGs are tissue specific. For example brevican, neurocan, neuroglycan C (all chondroitin sulfate or CSPGs) and phosphacan (CS/KSPG, a splice variant of the receptor-like protein tyrosine phosphatase-β) are mostly restricted to the central nervous (CNS) system [213, 214]. In the brain, PGs are found to be regulators of cell migration, axonal pathfinding, synaptogenesis and structural plasticity [215]. The cartilage also contains a variety of PGs like aggrecan, decorin, biglycan, fibromodulin and lumican that are essential for its normal performance. Cartilagenous proteoglycans are supposed to confer distinct textural properties that led to its classification into hyaline, elastic or fibrous type. Hyaline is the predominant cartilage form associated with the skeletal system whereas elastic cartilage is associated with the ear and the larynx. Fibrocartilage on the other hand is found in the menisci of the knee and inter-vertebral discs [216].

Proteoglycans (PGs) in the glomerular basement membrane (GBM) contribute to the renal function of glomerulus. Over 80% of PGs in the mature GBM are HSPG. Bamacan, a CS containing PG, however, is expressed transiently, only during nephrogenesis but disappears upon maturation [217]. In the developing kidney the HSPG perlecan is expressed very early, later it is accompanied by agrin. The core protein of perlecan, sized 467 kDa is composed of five structural domains [218]. Perlecan is widely expressed in basement membranes including those of skin, lung, colon, liver, heart, connective tissue, thymus, prostate, spleen, pituitary gland, kidney, placenta, skeletal muscle and blood vessels [219]. It can bind to various other molecules of the basement membrane, including nidogen, laminin (enhanced by nidogen), collagen IV and fibronectin. Self-association of perlecan is also reported [220]. Perlecan is widely known for tethering and modulation of the fibroblast growth factor (FBF) in the ECM. It can bind to FGF9 [221]; to FGF2 in cartilage [222] and the lens capsule of the eye [223] and also to FGF18 in cartilage [224]. It can also be associated with FGF10 in the BM of the sub-mandibular gland [225]. As per in vitro estimation one perlecan molecule can bind with a maximum of 123 molecules of FGF2 [224], though the number may be smaller in vivo as it interacts with many other ECM molecules.

Agrin, another HSPG is generally localized on the neuromuscular junction, where it plays a key role in the development of synaptic apparatus [226]. The non-neural isoforms of agrin is also found expressed in the kidney, lung and in the microvasculature [227,

228]. The large amount of homogenously distributed agrin in GBM led to the speculation that it might have an important role in the maintenance of glomerular permselectivity [229]. Alpha-dystroglycan is the cell surface receptor for agrin, which is expressed prominently in the epithelium of the developing and mature kidney. It localizes to the basal side of the epithelium where the cell surface is in contact with the basement membrane [230]. Interactions of agrin with integrins provide an additional structural link between the cytoskeleton and the ECM. The charge selective permeability of GBM is demonstrated to relate to the electrostatic properties of covalently bound heparan sulfates within the GBM [231, 232].

It is thus established that PGs serve diverse functions including ECM assembly and mediation of cell adhesion and motility [3, 233]. They associate with various secreted growth factors, enzymes and other functional proteins and modulate their half-life and effective availability to the surrounding cells. In a dynamic biological system this is achieved through (i) immobilization of functional sites; (ii) steric blocking of cleavage sites; (iii) delaying the active protein release through enzyme inhibition; (iv) protecting the functional protein from degradation and (v) masking and altering the concentration of subjected protein for effective presentation to cell surface receptors. The functional diversity of PGs can be attributed to the structural possibilities existing through the presence of multiple domains in the core protein as well as the variation in the number, position and chain length of the GAGs attached to it. The molecular diversity of decorin and aggrecan for example is the result of different combinations of protein core associated with one (in decorin) or more GAG chains of various subtypes (100 CS and 30 KS in aggrecan) [234, 235]. Similarly, perlecan though classified as HSPG, may also carry a CS chain, especially when expressed in cartilage. Enormous heterogeneity in the disaccharide branching of GAG chains in PGs makes them structurally complex molecules capable of bringing a degree/spectrum of interactivity with cells or other ECM partners [236]. An inherent range of interactive capabilities also exists due to the possible involvement of sulfate chains, protein cores or both.

2.1.3.9 Glycosaminoglycans (GAGs)

Glycosaminoglycans are polymers of repeating disaccharide units, the amino group of which may at times be acetylated. In ECM they exist predominantly as proteoglycans where they are covalently anchored to a core protein. The type and number of units of GAG largely determine the properties of PG [188]. Disaccharide unit in chondroitin sulfate (CS) and dermatan sulfate (DS) is galactosamine and glucuronic acid or iduronic acid, in heparin and in heparan sulfate (HS) it is glucosamine and either glucuronic acid or iduronic acid, and in keratin sulfate (KS) it is glucosamine and galactose (Fig. 2.11). The sulfated GAGs (CS, DS, and HS) are linked to their respective protein cores via O-glycosidic linkages at serine residues [186, 196]. All GAGs are negatively charged and have the propensity to attract ions. This creates an osmotic imbalance leading to absorption of water from their surroundings. The amount of GAGs in connective tissue is usually less than 10% by weight.

Tissues subjected to high compressive forces like articular cartilage have a large PG content (8–10% of the dry weight of the tissue) whereas tendons and ligaments which need to resist tension possess a very small concentration ($\sim$0.2% of dry weight of the tissue) [237]. At physiological pH, GAG chains contain one to three negative charges per disaccharide depending upon carboxylate and sulfate groups. The conspicuous hydration capacity of GAGs helps in filling the intercellular space while keeping multiple contacts within the fibrillar matrix. The multivalent branching of GAGs protects the core proteins of PGs and also the matrix-fibrils from proteolytic damage. Appropriately arranged GAGs in proteoglycans and glycoproteins are crucial for cell-ECM and cell-cell interactions. To a great extent tissue plasticity depends upon its GAG contents. They also assist in maintaining the viscoelasticity of blood vessels and the tensile strength of the skin and tendons.

2.1.3.10 Hyaluronan (hyaluronic acid)

Hyaluronate or hyluronan or hyluronic acid (HA) is a unique polymer with high water retention capacity (Fig. 2.12). It is often categorized as a glycosaminoglycan since it consists of repeating

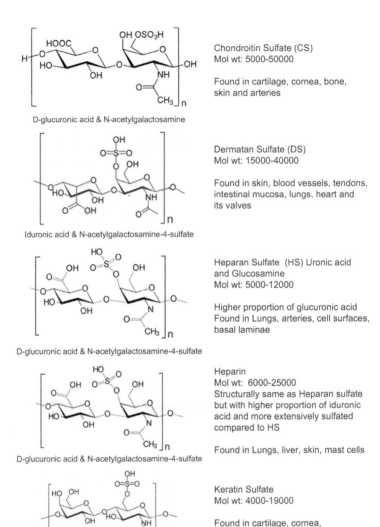

Chondroitin Sulfate (CS)
Mol wt: 5000-50000

Found in cartilage, cornea, bone, skin and arteries

D-glucuronic acid & N-acetylgalactosamine

Dermatan Sulfate (DS)
Mol wt: 15000-40000

Found in skin, blood vessels, tendons, intestinal mucosa, lungs, heart and its valves

Iduronic acid & N-acetylgalactosamine-4-sulfate

Heparan Sulfate (HS) Uronic acid and Glucosamine
Mol wt: 5000-12000

Higher proportion of glucuronic acid
Found in Lungs, arteries, cell surfaces, basal laminae

D-glucuronic acid & N-acetylgalactosamine-4-sulfate

Heparin
Mol wt: 6000-25000
Structurally same as Heparan sulfate but with higher proportion of iduronic acid and more extensively sulfated compared to HS

Found in Lungs, liver, skin, mast cells

D-glucuronic acid & N-acetylgalactosamine-4-sulfate

Keratin Sulfate
Mol wt: 4000-19000

Found in cartilage, cornea, inter-vertebral disc

D-galactose & N-acetyl glucosamine-6-sulfate

Figure 2.11   Common glycosaminoglycans (GAGs) of ECM.

disaccharide units but unlike other GAGs it is not sulfated [238]. Since it is not bound to a core protein it cannot be considered as a PG either. However, it is ubiquitously distributed in ECM and perhaps is the next most abundant ECM biopolymer after collagen.

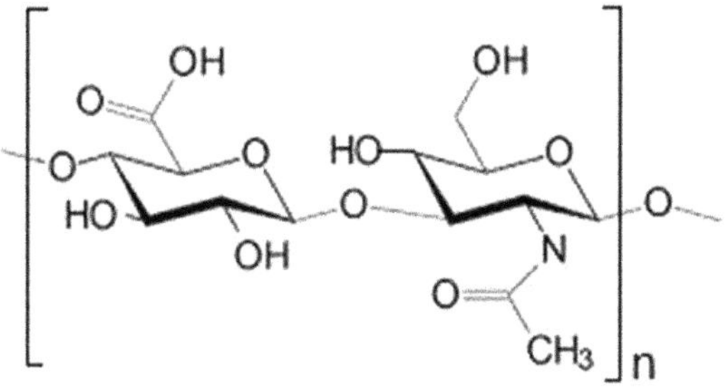

GlcA-β-1,3-GlcNAc-β-1,4-

Figure 2.12 Hyaluronic acid (glucuronic acid linked with N-acetyl glucosamine).

HA associates with collagens and other PGs to form supramolecular complexes. It readily imbibes a large amount of water which makes it a valuable ECM component of hydrated and soft tissues like tendon sheaths and bursae. HA is found in most of the connective tissues and is particularly concentrated in synovial fluid, the vitreous fluid of the eye, umbilical cord and chicken combs [237, 239]. Physiologically its role is attributed as a lubricant, helping protein fibers and other ECM components to slide over smoothly without much friction and damage. In cartilage where compressive forces are high, the proper distribution of HA is even more important. HA is naturally synthesized by a class of integral membrane enzymes called hyaluronan synthases and degraded by a family of enzymes called hyaluronidases [240].

In addition to collagens and proteoglycans, many non-collagenous glycoproteins also act as the building blocks of the ECM. The glycoprotein content of ECM is mostly related to the hydration and lubricating needs of a particular tissue. The role of glycosylated proteins and polysaccharides seems to be more of cushioning the cells from regular shear forces than scaffolding the ECM. Apparently they operate as adhesives while satisfying the cell's regular need of water and ions. It is realized that tissue with high collagen-fiber content and low amount of proteoglycans resist tensile forces whereas those with a high PG content well distributed within fibrous collagen network withstand compression [74].

Thus by examining the biopolymers that constitute ECM from different tissues it is clear that they impart morphological, textural and mechanical features to a tissue which in turn contribute to their physiological and mechanical functions. ECM constituents remain in terms/aligned with the cells on a continuous basis to keep the homeostatic balance. Any major metabolic change in the cell is reflected through ECM and any functional modulation in ECM is propopagated to the cells. The cell response in vivo is mainly influenced by cell-cell, cell-effector (signal) and/or cell-ECM interaction. These are the three major components/players of the tissue-specific microenvironment that dynamically prompt a cellular response.

2.2 Cell-Cell Interaction

Cell-cell communication and interaction is important for bringing about the functional co-ordination and co-operativity in a given tissue. Intercellular exchange of information may involve pre-existing contact junctions between the cells or may take place through transient contacts created by cell adhesion molecules (CAM). Cell-cell adhesion is a selective process and embryonic studies reveal that cells from one tissue (e.g., liver) specifically adhere to cells of the same tissue rather than to cells of a different tissue (e.g., brain). There are four different groups of CAM that are involved in cell-cell interaction; these are (i) integrins; (ii) selectins; (iii) cadherins and (iv) immunoglobulins (Ig). Except for members of the Ig super family, the other three adhesion mediators including integrins require Ca^{++} or Mg^{++} ions. Interactions through selectins, integrins and members of the Ig super family represent transient adhesions where cytoskeletons of adjacent cells are not linked to each other. Cadherins, the calcium dependent adherens on the other hand build stable adhesions where cytoskeletons of adhering cells remain connected. Thus, cadherin plays a role in the cell-cell interface similar to that played by integrin in cell-ECM interaction, i.e., it helps in connecting the cytoskeleton of the two interacting cells.

Cell-cell interaction and connection plays a key role in the organization of cells in tissues. Within a tissue the peripheral cells or group of cells may be permanently or stably connected through tight or gap junctions. Tight junctions are the closest known contacts between adjacent cells created by homophilic interactions of cloudins and occludins, the transmembrane proteins. Tight junctions often associated with cadheren junctions and desmosomes, are responsible for selective molecular trafficking and barrier like function of epithelium. They separate the apical and basolateral domains of the plasma membrane while sealing the free passage of molecules (including ions) between the cells of epithelial sheets. Gap junctions on the other hand provide an open channel through the plasma membrane allowing ions and small molecules to diffuse freely across the cytoplasm of adjacent cells. Gap junctions are shaped by connexin, another transmembrane protein. Six strands of connexin arrange as a cylinder which aligns with a similar assembly in the nearby cell to create an open channel connecting the cytoplasm of the two cells. Elevated calcium concentrations can close these channels. Communication through the gap junction is a common mechanism for metabolic coupling of adjoining cells. Most of the cells like epithelial, endothelial, cardiac and other smooth muscle cells in animal tissues communicate via gap junctions. Evidence exists for it being a prime mode of communication even in lymphocytes. These junctions also play an important role in achieving synchronized response in a tissue. Coordinated cellular action through apt intercellular communication is particularly crucial in the heart where rhythmic contraction is the key.

2.3 Cell-Effecter Interaction

Effectors are agents like hormones, growth factors or other soluble signaling molecules, which can influence the cell response. Being functionally significant communicators they can regulate different cell functions through autocrine (self), paracrine (adjacent) or endocrine (distant) signaling mechanisms. The endocrine signaling molecules generally function through specific receptors. They possess a strong affinity with their receptors which allows them to induce the effect even at low concentrations. Unlike endocrine signaling that involves blood for signal transportation to the target cells, autocrine and paracrine signaling is mediated through neighbouring ECM bio-polymers. Released effectors are generally diffused and presented to the surrounding cells through pericellular or extracellular matrix (ECM). Autocrine and juxtacrine signaling may also be mediated via membrane bound molecules for inducing itself (positive feedback) and/or neighboring cells. Such signals may also be sensed via cell surface receptors like integrins or may get entry into the cell through diffusion or osmotic passage across the cell membrane.

2.4 Cell-ECM Interaction

Years of research has proved that the growth and development of cells is greatly influenced by the extracellular matrix (ECM) present in immediate surroundings. Cells may contact and interact with ECM in a protrusive or contractile or mechanically supportive manner [74]. ECM biopolymers, especially the ones related to cell-matrix adhesion, are ancient and exquisitely conserved in multicellular animals [241]. Basement membranes are also commonly found in early invertebrates as well as in complex mammals including humans [9].

2.4.1 Protrusive Contacts

Protrusive contacts are made by the outward extension of plasma membrane. These protrusions are of a defined shape like filopodia, spikes or podosomes localized to a region of the cell. Currently these projections are distinguished from each other on the basis of actual lengths of projections as seen under the microscope. It is likely that they are part of the same dynamic intracellular structure that protrudes to create an extracellular contact. These projections may be broader in shape, forming pseudopodia, invadopodia or lamellopodia when non-localized. The adhesive receptors in the protrusive contacts are connected to the actin cytoskeleton but exert little tension on the matrix externally [242, 243]. These protrusions are known to mediate and guide the cells for matrix contact, adhesion and motility but how they differ in healthy, non-healthy (tumorous) and/or embryonic cells is not yet clear. Protrusive contacts are usually created during rapid membrane remodeling, transient matrix adhesions, exploratory cell movements and cell locomotion.

Bone remodeling exhibits an excellent example of transient matrix adhesions created by podosomes. It is understood that the mass and form of bone is regulated throughout life by the osteoblasts, which secrete and deposit bone ECM and osteoclasts, which resorb bone [244]. Both the osteoblasts and osteoclasts remain dynamically active and, depending upon the physiological forces and net stress, remodel the bone. In bones, osteoclasts exist in two functional states—the migratory state and the resorptive state. They show different types of cell–matrix contacts in both these states. Migratory osteoclasts use protrusive contacts for attachment. Through short cylindrical protrusions called podosomes they attach and move over the bone matrix. Podosomes are formed by the ventral cell membrane in which a core bundle of actin filament is connected to the plasma membrane by a circular array of cytoskeletal linker proteins [245, 246]. Matrix adhesions mediated through podosomes are highly transient and keep changing dynamically within 2–12 minutes during osteoclast motility. It has been shown that angiogenic endothelial cells in tumors also form podosome rosette [247]. This podosome rosette helps in degrading the ECM focally for vascular sprouting (Fig. 2.13).

Dystroglycan contacts are also protrusive cell-ECM contacts (Fig. 2.14). Dystroglycan is a glycoprotein which is composed of an alpha (120 kDa) and beta (43 kDa) subunits that are non-covalently associated. The alpha subunit is highly glycosylated extracellular glycoprotein whereas beta subunit is a transmembrane protein [248]. They are produced by the post-translation cleavage of precursor protein. Dystroglycan provides a link between laminin in the ECM and dystrophin in the membrane cytoskeleton. Glycosylation of the alpha-dystroglycan (156 kDa) allows it to link to laminin and the glycosylation of beta-dystroglycan (43 kDa) modulates its association with the intracellular cytoskeleton [249]. Defective glycosylation causes a group of muscular dystrophies called dystroglycanopathies, with a range of clinical severity. These

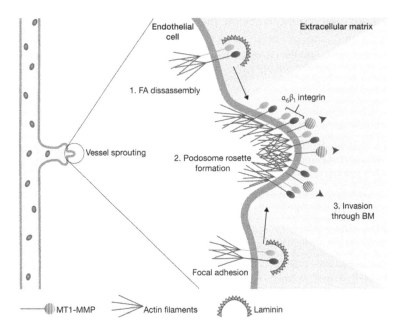

Figure 2.13 Schematic diagram of the steps involved in the formation of podosome rosettes during vessel sprouting. A mature vessel is shown on the left, with the area of the sprouting endothelial cell magnified on the right. Endothelial cell sprouting requires perforation of the basement membrane (BM) surrounding the mature vessel. (1) First, focal adhesions (FA) must disassemble, releasing $\alpha_6\beta_1$ integrin (green and purple) from its tight association with laminin (blue semicircles). (2) Next, focal adhesion components are relocalized (arrows) to form functional podosome rosettes (indicated by the arrows). (3) Activation of metalloproteinases, such as MT1-MMP (pink), within podosome rosettes allows the cell to invade (arrowheads) through the basement membrane, promoting the emergence of a new sprout. Actin filaments are shown in red, basement membrane is represented in light grey with the area of MT1-MMP-mediated extracellular matrix degradation shown in lighter grey, red ovals depict red blood cells inside the vessel. Figure adopted with permission from Ref. [247].

include congenital muscular dystrophies including Walker–Warburg syndrome, muscle-eye-brain disease.

Protrusive contacts are also implicated in cancer cell motility and invasion [250]. It has been reported that the progression of malignancy in cancer involves alterations in the expression profile of

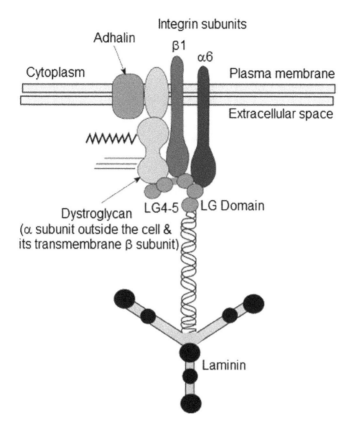

Figure 2.14 Contact of laminin LG domains 4 with dystroglycan and 2, 3 with integrins Ref. [248].

integrins and proteoglycans and down regulation of the fibronectin matrix. Cells exhibit decreased focal adhesions in culture and a switch to a more migratory phenotype that explains their invasive and metastatic behavior [251–254]. These changes in adhesion molecules and cell behavior lead to alterations in the predominant type of matrix contacts, from contractile focal adhesions and matrix assembly sites to protrusive contacts. Variations in the adhesion receptor profile can directly impact matrix interactions. In addition, the altered interplay between integrins, proteoglycans and other types of cell-surface receptors also exert potent effects on cancer cell migration.

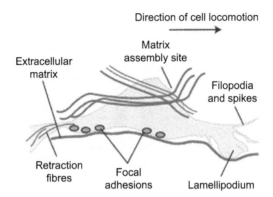

Figure 2.15 Schematic view of the matrix contacts formed by fibroblasts and other mesenchymal cells when migrating through extracellular matrix Ref. [250].

2.4.2 Contractile Contacts

Contractile contacts are relatively stronger where adhesive receptors make physically closer or longer-lasting contacts with the ECM. They are connected to the contractile actomyosin filament system inside the cell and as a consequence create isometric tension across the plasma membrane. Contractile contacts enable the cells to spread on a rigid surface or expand on deformable or flexible ECM scaffolds [255, 256]. Cell contacts through focal adhesions, collagen matrix assembly sites, fibrillar adhesions and fibrillin microfibril are some of the representative contractile contacts (Fig. 2.15).

2.4.3 Mechanically Supportive Contacts

Mechanically supportive contacts are the third category which include wider adhesions that can stabilize the plasma membrane and also enable the cells to resist deformation from mechanical forces. Cellular tensigrity which is mainly under the control of cytoskeleton represents the mechanically supportive contacts. The tensigrity model is first adopted by Ingber to explain the wide variations in cell phenotypes [257]. It also defines the mechanical status of the cell that is very much involved in the regulation of cell growth, migration and tissue patterning during morphogenesis.

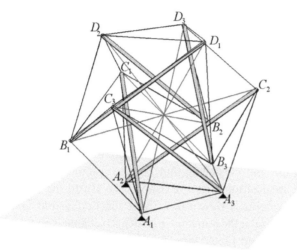

Figure 2.16 A spherical tensegrity structure with intermediate filaments used for generating the computational tensegrity model. The thin tendons represent microfilaments (black lines) and intermediate filaments (red lines); the thick gray struts indicate microtubules. Anchoring points to the substrate (blue) are indicated by the black triangles (A1, A2, and A3). Adopted from Refs. [254, 259].

Tensigrity theory and its translation into mathematical terms have opened up new avenues to correlate the cell mechanics with its biochemical behavior. This would certainly help in explaining the biological complexity of cell-ECM dynamics [258]. A computational tensigrity model (Fig. 2.16) that can predict dynamic rheological behaviors in living cells is already developed [259].

In general mechanically supportive contacts are made through cell surface receptors called Integrins [28, 260]. As the name suggests, these receptors integrate the cells with the surrounding ECM and act as a gateway to trigger many specific and non-specific molecular signaling pathways. Integrins are heterodimeric transmembrane molecules capable of triggering a cascade of intracellular events when they come in contact with specific molecules and/or ECM domains. The alpha and beta subunits of integrin are noncovalently associated. Each subunit has a large extracellular domain, a transmembrane segment, and a short intracellular tail of 35–50 residues, except β 4 whose intracellular domain is of approximately 1000 amino acids. Varients of alpha and beta subunits may dimerize and exist in different permutation combinations. Similar to basic ECM-constituents, integrins are also ancient molecules that evolved with time. In worms only two alpha and one beta subunits are involved in forming the two integrins whereas in mammals this number goes to more than 18 alpha and 8 beta subunits with over 24 combinations [27]. Compared to other specific receptors, integrins have lower binding affinity with their targets but that is compensated by their hundred-fold higher surface concentration and clustering that happens almost simultaneously with their association with the ligand. The clustering of integrins by matrix proteins is shown to activate focal adhesions which are mediated by focal adhesion kinase (FAK), a tyrosine kinase involved in phosphorylation [261]. Integrins are also capable of inside-out signaling [262, 263]. This possibility helps platelets when activated by injury to activate integrin binding to fibrinogen and forming a platelet clot.

Such cell surface receptor units can engage with different extracellular molecules while staying imbedded in the cell membrane bilayers. With their extracellular head region, most integrins bind extracellular matrix (ECM) macromolecules such as laminins and collagens in basement membranes or connective tissue components like fibronectin [264]. The ligand binding site of integrins is contributed by the extracellular domains of both α and β subunits, while the cytoplasmic domains interact directly or indirectly with several cytoskeletal-linker molecules and actin. Ligand binding concomitantly leads to integrin clustering, which is rapidly followed by recruitment of cytoskeletal linker molecules into adhesion complexes and the anchorage of actin stress fibers to these complexes. Thus, through biochemical interaction in extracellular space integrins can sense the environmental signal which is transmitted through a cascade of phosphorylation in the intracellular domain that rearranges the cytoskeleton [265]. In other words integrins act as switches that mechanically connect the extracellular matrix (ECM) to the cytoskeleton of the cell. The type and extent of interactive affinity impacts the cascade of events that follow the adhesion or engagement of integrin with ECM.

Hemidesmosomes that involve $\alpha6\beta4$ integrin are an excellent example of mechanically supportive contacts. Epithelial or endothelial cells attach to their underlying basement membrane via hemidesmosomes which links the non-contractile keratin filament system inside the cell [266, 267]. Matrix contacts through hemidesmosomes are specific to particular differentiated cell types while other mode of contacts such as spikes and filopodia may occur in many other cell types.

The most extensively studied motif present on multiple different domains of a number of ECM biopolymers is RGD (Arg-Gly-Glu). It interacts with the cells through $\alpha v\beta1$, $\alpha v\beta3$ or $\alpha IIb\beta3$ [268]. Collagen triple helix also interacts with the cells through integrins. However, since short collagen peptides unfold at physiological temperature, it has been difficult to map cell binding sites on collagen helices, except for the disulphide stabilized peptides. In collagen IV ($\alpha1\alpha1\alpha2$ trimer) two aspartate residues from each $\alpha1(IV)$ chain and one arginin residue from the $\alpha2(IV)$ chain were found to form a binding motif for $\alpha1\beta1$ integrin [269]. Unfolding of the triple helices abolishes the cell binding with $\alpha1\beta1$ and $\alpha2\beta1$ integrins and reveals new RGD-dependent binding [270–272]. It has also been realized that on denatured collagens only some of the RGD sequences are active and recognized by integrins. For example, among the 13 RGD sequences contributed by the human collagen VI alpha chains only one has activity and all the 3 RGD on collagen IV alpha chains are apparently inactive [272]. This suggests that the RGD motif is needed to be present in specific conformational orientation for its recognition by the inegrins. Structural analysis of RGD on fibronectin [273, 274] and tenacin [275] suggest a protruded and an open flap stereochemical presentation with both the acidic and basic chains pointing away in opposite directions.

It is understood that each cell can create multiple different types of matrix contacts at a given point of time. For example, other than podosomes as mentioned above, osteoclasts are also capable of making contacts with osteopontin, bone-sialoprotein and collagen within the bone matrix in an RGD dependent manner through $\alpha v\beta3$ integrin. The $\beta1$ integrins including $\alpha2\beta1$ on osteoclasts are also known to bind collagen I [276].

2.5 ECM Related Disorders

Developmental biology, gene alterations and knockout animal models have helped in revealing that some of the ECM components are indispensable for development and have innumerable roles other than providing structural support and imparting shape [3]. The role of ECM is diverse and sometimes governed exclusively by the type of tissue it belongs to. In connective tissues, for example, it provides mechanical support, tissue fluid transport, allows cell movement, and migration. It is becoming increasingly evident that the ECM extends control over different metabolic processes as well [74]. The tissue specificity of ECM is the guiding principle to uphold the normal development and physiological function. Mutations in ECM genes are known to cause a range of serious connective tissue disorders which may get inherited (Table 2.8). Tissues like bone, cartilage and cornea represent ECM-rich, structural tissues where impact of genetic mutations reflects more on morphology than the physiology of the tissue. Structural changes in collagen, the major ECM protein for example, are responsible for many of the ECM related diseases. On the other hand, cell-rich tissues where functionality is dependent on the orchestrated metabolic activity of cells, genetic mutations could be fatal. Unlike the ECM-rich, the cell dominant tissues like heart, lungs and CNS need peripheral ECM, only to retain the cell-cluster-morphology critical for acquiring functionality. ECM in these tissues is like glue that contributes to organize and bring physiological co-ordination among the adjoining cells.

The pathological impact of mutations or alterations in the ECM gene is manifested either through limited synthesis leading to reduced availability or misfolding of ECM proteins and glyco-proteins. Altered protein folding could cause undesired structural changes and consequent interactions. Misfolding of ECM molecules may also compromise the stability of the functional complexes in which it participates for certain physiological activities. Misfolded mutant proteins cause endoplasmic stress which is also suggested to contribute to the molecular pathology [277]. Since ECM is at the helm of integrating cells into functional assemblies, structural

Table 2.8 ECM components implicated in various diseases

ECM molecule	Genes involved	Tissue affected	Resultant disease
Collagen I	Col1α1, Col1α2	Bone	Osteogenesis imperfecta[a]
	Col1α1, Col1α2	Skin joints	Ehler-Danlos syndrome, type VII[b]
	Col1α2	Skin joints, heart	Ehler-Danlos syndrome, cardiovalvualar form[c]
Collagen II	Col2α1	Cartilage, eyes	Spondyloepiphyseal dysplasia[c] Spondyloepimetaphyseal dysplasia[c] Achondrogenesis, hypochondrogenesis Kniest dysplasia, Stickler syndrome
Collagen III	Col3α1	Blood vessels	Ehler-Danlos syndrome, type VII[b,c]
Collagen IV	Col4α1	Kidney, skin, BM	Familial porencephaly, hereditary angiopathy with nephropathy, aneurysma & muscle cramps syndrome[d]
	Col4α3, Col4α4	Kidney, skin, BM	Alport syndrome, Benign familial Haematuria[d,e]
	Col4α5, Col4α6	Kidney, skin, BM	Alport syndrome, Leiomyomatosis[d–f]
Collagen V	Col5α1, Col5α2	Skin, joints	Ehler-Danlos syndrome, type I & II[g]
Collagen VI	Col6α1, Col6α2 Col6α3	Muscle	Bethlem myopathy, Ullrich congenital muscular dystrophy[h]
Collagen VII	Col7α1	Skin	Dystrophic epidermolysis bullosa[i] (dermal-epidermal junction)
Collagen VIII	Col8α1	Cornea	Fuchs corneal dystrophy
	Col8α2	Cornea	Corneal endothelial dystrophies[f]

(Continued)

Table 2.8 (*Continued*)

ECM molecule	Genes involved	Tissue affected	Resultant disease
Collagen IX	Col9α1, Col9α2 Col9α3	Cartilage	Multiple epiphyseal dysplasia[j]
	Col9α1	Cartilage	Autosomal recessive Stickler syndrome[j]
Collagen X	Col10α1	Cartilage, growth plate	Metaphyseal chondrodysplasia, Schmid type[k]
Collagen XI	Col11α1, Col11α2	Cartilage, eyes	Stickler syndrome, Marshall syndrome[j]
	Col11α2	Cartilage, ears	Otospondylomegaepiphyseal dysplasia[j]
	Col11α2	Ears	Deafness[j]
Collagen XVII	Col17α1	Skin	Junctional epidermolysis bullosa[l]
Collagen XVIII	Col18α1	Eyes, skull	Knobloch syndrome, type I[m]
Aggrecan	ACAN	Cartilage	Spondyloepiphyseal dysplasia, Kimberley type
Thrombospondin	COMP*	Cartilage, ligaments	Multiple epiphyseal dyaplasia[j]
Decorin	DCN	Cornea	Congenital stromal corneal dystrophy[n]
Elastin	ELN	Arteries, skin	Supravalvular aortic stenosis, cutis lasca[n]
Fibronectin	FN1	Kidney	Glomerulopathy[n]
Laminin	LAMA2	Muscle	Congenital muscular dystrophy[n]
	LAMA3, LAMB3, LAMC2	Skin	Epidermolysis bullosa junctional[n] (dermal-epidermal junction)
Fibrillin1	FBN1	Skeleton, eyes cardiovascular	Ectopia lentis, Marfan syndrome, Shprintzen Goldberg syndrome, Weill-Marchesani syndrome[n]

(*Continued*)

Table 2.8 (*Continued*)

ECM molecule	Genes involved	Tissue affected	Resultant disease
Fibrillin2	FBN2	Skeleton	Contractural arachnodactyly[n]
Fibulin4	FBLN4	Skin	Cutis laxa[n]
Fibulin5	FBLN5	Eyes	Age-related macular degeneration[n]
Fibulin5	FBLN5	Skin	Cutis laxa[14]
Matrillin3	MATN3	Cartilage	Multiple epiphyseal dysplasia[n]
Perlecan	HSPG2	Cartilage (basement membrane)	Schwartz-Jampel syndrome, dysegmental dysplasia, Silverman-Handmaker type[n]
TenascinXB	TNXB	Skin	Ehlers-Danlos-like syndrome[n]
	TNXB	Skin	Ehlers-Danlos syndrome, type III[n]

*Cartilage oligomeric matrix protein also known as thrombospondin 5.

Table references:

[a]Marini et al. 2007, *Hum Mutat*, **28**, 209; [b]Dalgleish R 1998, *Nucleic Acids Res*, **26**, 253; [c]Bodian & Klein 2009, *Hum Mutat*, **30**, 946; [d]Van Agtmael & Bruckner-Tuderman 2010, *Cell Tissue Res*, **339**, 167; [e]Bruckner-Tuderman et al. 1987, *Eur J Biochem*, **165**, 607; [f]Bateman et al. 2009, *Nat Rev Genet*, **10**, 173 (Ref. [277]); [g]Callewaert et al. 2008, *Best Pract Res Clin Rheumatol*, **22**, 165; [h]Lampe & Bushby 2005, *J Med Genet*, **4**, 673; [i]Fine JD 2010, *Curr Opin Pediatr*, **22**, 453; [j]Carter & Raggio 2009, *Curr Opin Pediatr*, **21**, 46; [k]Grant ME 2007, *Int J Exp Pathol*, **88**, 203; [l]Has & Kern 2010, *Dermatol Clin*, **28**, 61; [m]Nicolae & Olsen 2010, *Cell Tissue Res*, **339**, 155.

Table 2.9 Diseases associated with abnormal ECM or cell-ECM contacts

ECM abnormality	Disease/pathology	Nature of disease	Cause
Dystroglycan contacts	Muscular dystrophies	Acquired	Mutations of Laminin
	Dilated cardiomyopathy	Inherited	
	Leprosy	Acquired	*Mycobacterium leprae* binds α-dystroglycan
	Lassa fever	Acquired	Arenavirus binds α-dystroglycan
	Enteroviral cardiomyopathy	Acquired	Cleavage of dystrophin by Coxsaclevirus protease A2
Fibrillar adhesion (reduced)	Cancer	Acquired	Altered Integrin expression
			Decreased fibronectin matrix assembly
Fibrillin microfibril	Marfan's syndrome, Cutis laxa, rupture of blood vessels	Acquired	Mutations of Fibrillin 1
		Acquired	Mutations of tropoelastin
Focal adhesion	Von Willebrand's disease (altered platelet attachments)	Acquired	Mutations of Von Willebrand factor
		Inherited	Absence of Von Willebrand factor
	Bernard Souller syndrome	Acquired	Mutations in platelet glycoprotein 1b
		Inherited	Absence of platelet glycoprotein 1b
	Glanzmann's thrombasthenia	Acquired	Mutations in $\alpha 1\alpha 3$ integrin that blocks platelet adhesion
Hemidesmosome	Epidermolysis bullosas	Acquired/ inherited	Mutation or absence of laminin$\alpha 6$ or $\alpha 4$ integrin subunits, plectin, BPAG1e or keratins 5 or 14 that destabilizes contacts
Invadopodium with cell motility	Invasive cancers	Acquired	Upregulation of contacts associated
Spike/microspike	Invasive carcinoma	Acquired	Upregulation of fascin correlates with Invasiveness; Association of $\alpha 6\alpha 4$ integrin with actin; Protrusions associated with cell motility

alterations, especially in the regions involved in making connections may also prove pathological.

Table 2.9 represents some of the disorders caused by mutations in the genes that code for collagen α chains. Data for the above table is adopted from Ref. [250].

2.6 Conclusion

2.6.1 Classification

It is clear now that most of the ECM molecules are chimeric in nature and may possess common or similar structural or functional domains. This has made their classification tedious. For example, some collagens are dominated by their characteristic triple helical structures while others have only intermittent stretches of it. Collagens IX, XII and XIV are constituted of GAGs and of non-collagenous domains that correspond to almost 90% of their molecular mass. They generally are differentially spliced leading to changes in primary sequences. Furthermore, the collagen triple helix that is considered typical of ECM polymers is also found in other molecules like the macrophage scavenger receptor, the acetyl-cholinesterase receptor, and the C1q component of complement [197]. Therefore time is right for classifying ECM molecules into categories like indispensable, crucial (essential) and supplementary (non-essential) yet assistive on the basis of deleterious consequences exhibited by the functional and non-functional cell/tissue phenotypes obtained after specific deletions since some deletions get compensated by other functionally robust biopolymers. For example, Lysyl oxidase (LOX) and LOX-like (LOXL1) protein cross-link both elastin and collagen but mice, lacking the LOX gene (lox-/-), die at the end of gestation or just after birth whereas mice lacking the LOXL gene (loxl1-/-) live a normal lifespan and do not show the obvious vascular and pulmonary defects apparent in lox-/- mice. Creating a database of such information would be useful in scrutinizing critical, disease causing genetic mutations in ECM related biopolymers.

2.6.2 Inter & Intramolecular ECM Interactions

The interplay among the known ECM components is complex and baffling. Presence of common functional domains in different ECM polymers and multiple copies or repeat functional motives and domains that sometime remain encrypted within the tertiary structure of the ECM molecule are also found very frequently. Triggering pathways to make these encrypted sequences physiologically available as and when the need arises and only few, not all, the motives are active at times, is intriguing. Probably this differential activity of motives is governed by their immediate environment or neighbouring sequence that restrict or limit their molecular expansion and thus conformational availability to the receptor. The regulation of the functional ECM-motif interaction at the site could also be exercised from the cell surface receptor region. The alignment of alpha, beta chains of Integrins and their clustering as per the intensity of the extracellular signals is part of cell-ECM dynamics' regulation. It is likely that multiple copies of functional domains through clustering satisfy the need while the other end of the unfolded molecule is capable of concommitted functions like providing the feedback to the cell to trigger the expression of more receptors to meet the demand. Thus, other than functional modulation of receptor sites at the interface of the cell and ECM, the folding and unfolding of cryptic functional domains of ECM polymers seems to be the most common mechanism that keeps cells and the surrounding ECM in tandem. Our understanding of the molecular basis of ECM biomolecules and their functions vis-à-vis cells is expanding rapidly through novel tools and techniques. They could be complemented by genetic and molecular studies on human diseases and transgenic mice with deleted, mutated or truncated genes. However, it would be worthwhile to determine the role of the most common ECM constituents to understand the fundamentals of their impact on cell response. Functional domains and sequences that are common in different ECM constituents need to be characterized and catalogued for further reference. This would help comprehending ECM-cell dynamics in a specific scenario.

2.6.3 Implications

The significance of ECM for the existence and survival of cells, tissue and, in turn, organism is now well established. Genetic disorders and their roots to ECM mutations further emphasize the integrated role of ECM genes and their products [277]. Structural aberrations present in hemidesmosomes, the epidermal-dermal junctions which help in linking epidermis to the basement membrane is linked with an autoimmune disease called cutaneous blistering. The bullous pemphigoid antigen causing the condition is found to be a very special collagen XVII. It is present as a transmembrane collagen with type II orientation, i.e., the carboxyterminal portion being extracellular [278–280].

References

1. Burridge K, Fath K, Kelly T, Nuckolls G and Turner C (1988). Focal adhesions: transmembrane junctions between the extracellular matrix and the cytoskeleton. *Ann Rev Cell Biol*, **4**, 487–525.

2. Vogt G, Huber M, Hiemann M, van den Boogaart G, Schmitz OJ and Schubart CD (2008). Production of different phenotypes from the same genotype in the same environment by developmental variation. *J Exp Biol*, **211**, 510–523.

3. Tsang KY, Cheung MC, Chan D and Cheah KS (2010). The developmental roles of the extracellular matrix: beyond structure to regulation. *Cell Tissue Res*, **339**, 93–110.

4. Aszodi A, Legate KR, Nakchbandi I and Fassler R (2006). What mouse mutants teach us about extracellular matrix function. *Annu Rev Cell Dev Biol*, **22**, 591–621.

5. Frantz C, Stewart KM and Weaver VM (2010). The extracellular matrix at a glance. *J Cell Sci*, **123**, 4195–4200.

6. Kelleher CM, McLean SE and Mecham RP (2004). Chapter 6 - Vascular extracellular matrix and aortic development, in *Current Topics in Developmental Biology*, Vol. 62, pp. 153–188.

7. Ayad S, Boot-Handford RP, Humphries MJ, Kadler KE and Adrian Shuttleworth C (1998). *The Extracellular Matrix FactsBook*, 2nd edn. FactsBook Series, Academic Press.

8. Hynes RO (2009). The extracellular matrix: not just pretty fibrils. *Science*, **326**, 1216–1219.

9. Rhodes JM and Simons M (2007). The extracellular matrix and blood vessel formation: not just a scaffold. *J Cell Mol Med*, **11**(2), 176–205.

10. Guilak F, Alexopoulos LG, Upton ML, Youn I, Choi JB, Cao L, et al. (2006). The pericellular matrix as a transducer of biomechanical and biochemical signals in articular cartilage. *Ann N Y Acad Sci*, **1068**, 498–512.

11. Engvall E, Hessle H and Klier G (1986). Molecular assembly, secretion, and matrix deposition of type-VI collagen. *J Cell Biol*, **102**, 703–710.

12. Poole CA, Flint MH and Beaumont BW (1987). Chondrons in cartilage - ultrastructural analysis of the pericellular microenvironment in adult human articular cartilages. *J Orthop Res*, **5**, 509–522.

13. Ruoslahti E and Yamaguchi Y (1991). Proteoglycans as modulators of growth-factor activities. *Cell*, **64**, 867–869.

14. Nicodemus GD, Skaalure SC and Bryant SJ (2011). Gel structure impacts pericellular and extracellular matrix deposition which subsequently alters metabolic activities in chondrocyte-laden PEG hydrogels. *Acta Biomater*, **7**(2), 492–504.

15. Lukashev ME and Werb Z (1998). ECM signalling: orchestrating cell behaviour and misbehaviour. *Trends Cell Biol*, **8**, 437–441.

16. Ramirez F and Rifkin DB (2003). Cell signaling events: a view from the matrix. *Matrix Biol*, **22**, 101–107.

17. Ramirez F and Rifkin DB (2009). Extracellular microfibrils: contextual platforms for TGF-beta and BMP signaling. *Curr Opin Cell Biol*, **21**, 616–622.

18. Boudreau N and Weaver V (2006). Forcing the third dimension. *Cell*, **125**, 429–431.

19. Daley WP, Peters SB and Larsen M (2008). Extracellular matrix dynamics in development and regenerative medicine. *J Cell Sci*, **121**, 255–264.

20. Roy R, Zhang B and Moses MA (2006). Making the cut: protease-mediated regulation of angiogenesis. *Exp Cell Res*, **312**, 608–622.

21. van Hinsbergh VW, Engelse MA and Quax PH (2006). Pericellular proteases in angiogenesis and vasculogenesis. *Arterioscler Thromb Vasc Biol*, **26**, 716–778.

22. Koblinski JE, Ahram M and Sloane BF (2000). Unraveling the role of proteases in cancer. *Clin Chim Acta*, **291**, 113–135.

23. Mohamed MM and Sloane BF (2006). Cysteine cathepsins: multifunctional enzymes in cancer. *Nat Rev Cancer*, **6**, 764–775.

24. Dutta RC and Dutta AK (2011). ECM analog technology for 3D cell culture. *Front Biosci (Elite Ed)*, **4**, 1043–1048.

25. Bianco A, Poukkula M, Cliffe A, Mathieu J, Luque CM, Fulga TA and Rørth P (2007). Two distinct modes of guidance signalling during collective migration of border cells. *Nat Lett*, **448**, 362–366.

26. Takada Y, Ye X and Simon S (2007). The integrins. *Genome Biol*, **8**(5), 215.

27. Humphries JD, Byron A and Humphries MJ (2006). Integrin ligands at a glance. *J Cell Sci*, **119**, 3901–3903.

28. Hynes RO (2002). Integrins: bidirectional, allosteric signaling machines. *Cell*, **110**, 673–687.

29. Vogel V and Baneyx G (2003). The tissue engineering puzzle: a molecular perspective. *Annu Rev Biomed Eng*, **5**, 441–463.

30. Hallmann R, Horn N, Selg M, Wendler O, Pausch F and Sorokin LM (2005). Expression and function of laminins in the embryonic and mature vasculature. *Physiol Rev*, **85**, 979–1000.

31. LeBleu VS, Macdonald B and Kalluri R (2007). Structure and function of basement membranes. *Exp Biol Med*, **232**, 1121–1129.

32. Huxley-Jones J, Robertson DL and Boot-Handford RP (2007). On the origins of the extracellular matrix in vertebrates. *Matrix Biol*, **26**, 2–11.

33. Bhaskaran M, Kolliputi N, Wang Y, Gou D, Chintagari NR and Liu L (2007). Trans-differentiation of alveolar epithelial type II Cells to type I cells involves autocrine signaling by transforming growth factor beta 1 through the smad pathway. *J Biol Chem*, **282**, 3968–3976.

34. Shenoi BV (2004). In vitro lymphocyte-to-granulocyte transdifferentiation induced by chemicals. *Curr Sci*, **87**, 491–494.

35. Mecham RP (1998). Overview of extracellular matrix. *Curr Protoc Cell Biol*, 10.1.1–10.1.14.

36. Ricard-Blum S (2011). The collagen family. *Cold Spring Harb Perspect Biol*, **3**(1), a004978.

37. Ricard-Blum S and Ruggiero F (2005). The collagen superfamily: from the extracellular matrix to the cell membrane. *Pathol Biol*, **53**, 430–442.

38. Gordon MK and Hahn RA (2010). Collagens. *Cell Tissue Res*, **339**, 247–257.

39. Oh SP, Kamagata Y, Muragaki Y, Timmons S, Ooshima A and Olsen BR (1994). Isolation and sequencing of cDNAs for proteins with multiple domains of Gly-Xaa-Yaa repeats identify a distinct family of collagenous proteins. *Proc Natl Acad Sci USA*, **91**, 4229–4233.

40. Kielty CM, Hopkinson I and Grant ME (1993). Collagen: the collagen family, structure, assembly, and organization in the extracellular matix. In: Royce PM and Steinmann BS, eds. *Connective Tissue and Its Heritable Disorders: Molecular, Genetic, and Medical Aspects.* New York, NY: Wiley-Liss, pp. 103–147.

41. Myllyharju J and Kivirikko KI (2004). Collagens, modifying enzymes and their mutations in humans, flies and worms. *Trends Genet*, **20**, 33–43.

42. Greenspan DS (2005). Biosynthetic processing of collagen molecules. *Top Curr Chem*, **247**, 149–183.

43. Mäki JM (2009). Lysyl oxidases in mammalian development and certain pathological conditions. *Histol Histopathol*, **24**, 651–660.

44. Nagata K (2003). HSP47 as a collagen-specific molecular chaperone: function and expression in normal mouse development. *Semin Cell Dev Biol*, **14**, 275–282.

45. Esposito C and Caputo I (2005). Mammalian transglutaminases identification of substrates as a key to physiological function and physiopathological relevance. *FEBS J*, **272**, 615–631.

46. Avery NC and Bailey AJ (2006). The effects of the Maillard reaction on the physical properties and cell interactions of collagen. *Pathol Biol*, **54**, 387–395.

47. Vanacore R, Ham AJ, Voehler M, Sanders CR, Conrads TP, Veenstra TD, Sharpless KB, Dawson PE and Hudson BG (2009). A sulfilimine bond identified in collagen IV. *Science*, **325**, 1230–1234.

48. Eyre DR and Wu JJ (2005). Collagen cross-links. *Top Curr Chem*, **247**, 207–230.

49. Wu JJ, Weis MA, Kim LS, Carter BG and Eyre DR (2009). Differences in chain usage and cross-linking specificities of cartilage type V/XI collagen isoforms with age and tissue. *J Biol Chem*, **284**, 5539–5545.

50. Wu JJ, Weis MA, Kim LS and Eyre DR (2010). Type III collagen, a fibril network modifier in articular cartilage. *J Biol Chem*, **285**, 18537–18544.

51. Eyre DR, Weis MA and Wu JJ (2010). Maturation of collagen ketoimine cross-links by an alternative mechanism to pyridinoline formation in cartilage. *J Biol Chem*, **285**, 16675–16682.

52. Saito M and Marumo K (2010). Collagen cross-links as a determinant of bone quality: a possible explanation for bone fragility in aging, osteoporosis and diabetes mellitus. *Osteoporos Int*, **21**, 195–214.

53. Sjöberg JS and Bulterijs S (2009). Characteristics, formation, and pathophysiology of glucosepane: a major protein cross-link. *Rejuvenation Res*, **12**, 137–148.

54. Kadler KE, Baldock C, Bella J and Boot-Handford RP (2007). Collagens at a glance. *J Cell Sci*, **120**, 1955–1958.

55. Sabeh F, Ota I, Holmbeck K, Birkedal-Hansen H, Soloway P, Balbin M, Lopez-Otin C, Shapiro S, Inada M, Krane S, et al. (2004). Tumor cell traffic through the extracellular matrix is controlled by the membrane-anchored collagenase MT1-MMP. *J Cell Biol*, **167**, 769–781.

56. Linsenmayer TF (1991). Collagen. In: Hay ED, ed. *Cell Biology of Extracellular Matrix*. New York, NY: Plenum Press, pp. 7–44.

57. Mayne R (1986). Collagenous proteins of blood vessels. *Arteriosclerosis*, **6**, 585–593.

58. Burgeson RE and Nimni ME (1992). Collagen types: molecular structure and tissue distribution. *Clin Orthop*, **282**, 250–272.

59. Clark RAF (1985). Cutaneous tissue repair, I: basic biologic considerations. *J Am Acad Dermatol*, **13**, 701–725.

60. Flint MH (1990). Connective tissue biology. In: McFarlane RM, McCrouther DA, Flint MH, eds. *Dupuytren's Disease*. Edinburgh, Scotland: Churchill Livingstone, pp. 13–24.

61. Kuhn K, Wiedemann H, Timpl R, Risteli J, Dieringer H, Voss T and Glanville RW (1981). Macromolecular structure of basement membrane collagens. *FEBS Lett*, **125**, 123–128.

62. Timpl R, Wiedemann H, van Delden V, Furthmayr H and Kuhn K (1981). A network model for the organization of type IV collagen molecules in basement membranes. *Eur J Biochem*, **120**, 203–211.

63. Poschl E, Schlotzer-Schrehardt U, Brachvogel B, Saito K, Ninomiya Y and Mayer U (2004). Collagen IV is essential for basement membrane stability but dispensable for initiation of its assembly during early development. *Development*, **131**, 1619–1628.

64. Timpl R and Brown JC (1996). Supramolecular assembly of basement membranes. *Bioessays*, **18**, 123–132.

65. Marchant JK, Hahn RA, Linsenmayer TF and Birk DE (1996). Reduction of type V collagen using a dominant-negative strategy alters the regulation of fibrillogenesis and results in the loss of corneal-specific fibril morphology. *J Cell Biol*, **135**, 1415–1426.

66. Christiano AM, Greenspan DS, Lee S and Uitto J (1994). Cloning of human type VII collagen. Complete primary sequence of the alpha 1(VII) chain and identification of intragenic polymorphisms. *J Biol Chem*, **269**, 20256–20262.

67. Uitto J, Chung-Honet LC and Christiano AM (1992). Molecular biology and pathology of type VII collagen. *Exp Dermatol*, **1**, 2–11.

68. Burgeson RE (1993). Type VII collagen, anchoring fibrils, and epidermolysis bullosa. *J Invest Dermatol*, **101**, 252–255.

69. Esposito C and Caputo I (2005). Mammalian transglutaminases identification of substrates as a key to physiological function and physiopathological relevance. *FEBS J*, **272**, 615–631.

70. Vanacore R, Ham AJ, Voehler M, Sanders CR, Conrads TP, Veenstra TD, Sharpless KB, Dawson PE and Hudson BG (2009). A sulfilimine bond identified in collagen IV. *Science*, **325**, 1230–1234.

71. Jozsa L, Kannus P, Balint JB and Reffy A (1991). Three-dimensional ultrastructure of human tendons. *Acta Anat (Basel)*, **142**, 306–312.

72. Harkness RD (1980). Mechanical properties of connective tissues in relation to function. In: Parry DAD and Creamer LK, eds. *Fibrous Proteins: Scientific, Industrial, and Medical Aspects*. London, England: Academic Press, pp. 207–230.

73. Liu SH, Yang R-S, al-Shaikh R and Lane JM (1995). Collagen in tendon, ligament, and bone healing: a current review. *Clin Orthop*, **318**, 265–278.

74. Culav EM, Clark CH and Merrilees MJ (1999). Connective tissues: matrix composition and its relevance to physical therapy. *Phys Ther*, **79**, 308–319.

75. Nicolae C and Olsen BR (2010). Unexpected matrix diseases and novel therapeutic strategies. *Cell Tissue Res*, **339**(1):155–165.

76. Fraser DA and Tenner AJ (2008). Directing an appropriate immune response: the role of defense collagens and other soluble pattern recognition molecules. *Curr Drug Targets*, **9**, 113–122.

77. Maertens B, Hopkins D, Franzke CW, Keene DR, Bruckner-Tuderman L, Greenspan DS and Koch M (2007). Cleavage and oligomerization of gliomedin, a transmembrane collagen required for node of Ranvier formation. *J Biol Chem*, **282**, 10647–10659.

78. Heino J (2007). The collagen family members as cell adhesion proteins. *Bioessays*, **29**, 1001–1010.

79. Ricard-Blum S and Ballut L (2011). Matricryptins from collagens and proteoglycans. *Front Biosci*, **16**, 674–697.

80. Brooke BS, Bayes-Genis A and Li DY (2003). New insights into elastin and vascular disease. *Trends Cardiovasc Med*, **13**, 176–181.

81. Elastin RJ (1993). In: Royce PM and Steinmann BS, eds. *Connective Tissue and Its Heritable Disorders: Molecular, Genetic, and Medical Aspects*. New York, NY: Wiley-Liss, pp. 167–188.

82. Hirano E, Knutsen RH, Sugitani H, et al. (2007). Functional rescue of elastin insufficiency in mice by the human elastin gene. implications for mouse models of human disease. *Circ Res*, **101**, 523–531.

83. Rosenbloom J, Abrams WR and Mecham R (1993). Extracellular matrix 4: the elastic fiber. *FASEB J*, **7**, 1208–1218.

84. Chambers RC and Laurent GJ (1996). The lung. In: Comper WD, ed. *Extracellular Matrix*, Vol. 1: Tissue Function. Amsterdam, the Netherlands: Harwood Academic Publishers, pp. 378–409.

85. Sandberg LB, Soskel NT and Leslie JG (1981). Elastin structure, biosynthesis, and relation to disease states. *N Engl J Med*, **304**, 566–579.

86. Flint MH (1990). Connective tissue biology. In: McFarlane RM, McCrouther DA and Flint MH, eds. *Dupuytren's Disease*. Edinburgh, Scotland: Churchill Livingstone, pp. 13–24.

87. Bernstein EF, Chen YQ, Tamai K, et al. (1994). Enhanced elastin and fibrillin gene expression in chronically photodamaged skin. *J Invest Dermatol*, **103**, 182–186.

88. Partridge SM, Elsden DF and Thomas J (1963). Constitution of the cross-linkages in elastin. *Nature*, **197**, 1297–1298.

89. Partridge SM, Elsden DF, Thomas J, et al. (1964). Biosynthesis of the desmosine and isodesmosine crossbridges in elastin. *Biochem J*, **93**, 30C–33C.

90. Wagenseil JE and Mecham RP (2007). New insights into elastic fiber assembly. *Birth Defects Res C*, **81**, 229–240.

91. Davis EC, Roth RA, Heuser JE and Mecham RP (2002). Ultrastructural properties of ciliary zonule microfibrils. *J Struct Biol*, **139**, 65–75.

92. Gibson MA, Sandberg LB, Grosso LE and Cleary EG (1991). Complementary DNA cloning establishes microfibril-associated glycoprotein (MAGP) to be a discrete component of the elastin-associated microfibrils. *J Biol Chem*, **266**, 7596–7601.

93. Gibson MA, Hatzinikolas G, Kumaratilake JS, Sandberg LB, Nicholl JK, Sutherland GR and Cleary EG (1996). Further characterization of proteins associated with elastic fiber microfibrils including the molecular cloning of MAGP-2 (MP25). *J Biol Chem*, **271**, 1096–1103.

94. Trask BC, Broekelmann T, Ritty TM, Trask TM, Tisdale C and Mecham RP (2001). Posttranslational modifications of microfibril associated glycoprotein-1 (MAGP-1). *Biochemistry*, **40**, 4372–4380.

95. Frankfater C, Maus E, Gaal K, Segade F, Copeland NG, Gilbert DJ, Jenkins NA and Shipley JM (2000). Organization of the mouse microfibril-associated glycoprotein-2 (MAGP-2) gene. *Mam Genom*, **11**, 191–195.

96. Gibson MA, Finnis ML, Kumaratilake JS and Cleary EG (1998). Microfibril-associated glycoprotein-2 (MAGP-2) is specifically associated with fibrillin-containing microfibrils but exhibits more restricted patterns of tissue localization and developmental expression than its structural relative MAGP-1. *J Histochm Cytochem*, **46**, 871–885.

97. Weinbaum JS, Broekelmann TJ, Pierce RA, Werneck CC, Segade F, Craft CS, Knutsen RH and Mecham RP (2008). Deficiency in microfibril-associated glycoprotein-1 leads to complex phenotypes in multiple organ systems. *J Biol Chem*, **283**, 25533–25543.

98. Downing AK, Knott V, Werner JM, et al. (1996). Solution structure of a pair of calcium-binding epidermal growth factor-like domains: implications for the Marfan syndrome and other genetic disorders. *Cell*, **85**, 597–605.

99. Corson GM, Charbonneau NL, Keene DR and Sakai LY (2004). Differential expression of fibrillin-3 adds to microfibril variety in human and avian, but not rodent, connective tissues. *Genomics*, **83**, 461–472.

100. Handford PA, Downing AK, Reinhardt DP and Sakai LY (2000). Fibrillin: from domain structure to supramolecular assembly. *Matrix Biol*, **19**, 457–470.

101. Trask TM, Ritty TM, Broekelmann T, Tisdale C and Mecham RP (1999). N-terminal domains of fibrillin 1 and fibrillin 2 direct the formation of homodimers: a possible first step in microfibril assembly. *Biochem J*, **340**, 693–701.

102. Trask TM, Trask BC, Ritty TM, Abrams WR, Rosenbloom J and Mecham RP (2000). Interaction of tropoelastin with the amino-terminal domains of fibrillin-1 and fibrillin-2 suggests a role for the fibrillins in elastic fiber assembly. *J Biol Chem*, **275**(32), 24400–24406.

103. Neptune ER, Frischmeyer PA, Arking DE, Myers L, Bunton TE, Gayraud B, Ramirez F, Sakai LY and Dietz HC (2003). Dysregulation of TGF-beta activation contributes to pathogenesis in Marfan syndrome. *Nat Genet*, **33**, 407–411.

104. Ritty TM, Broekelmann TJ, Werneck CC and Mecham RP (2003). Fibrillin-1 and -2 contain heparin-binding sites important for matrix deposition and that support cell attachment. *Biochem J*, **375**, 425–432.

105. Carta L, Pereira L, Arteaga-Solis E, et al. (2006). Fibrillins 1 and 2 perform partially overlapping functions during aortic development. *J Biol Chem*, **281**, 8016–8023.

106. Argraves WS, Greene LM, Cooley MA and Gallagher WM (2003). Fibulins: physiological and disease perspectives. *EMBO Rep*, **4**, 1127–1131.

107. Chu ML and Tsuda T (2004). Fibulins in development and heritable disease. *Birth Defects Res C Embryo Today*, **72**, 25–36.

108. Kobayashi N, Kostka G, Garbe JH, Keene DR, Bachinger HP, Hanisch FG, Markova D, Tsuda T, Timpl R, Chu ML and Sasaki T (2007). A comparative analysis of the fibulin protein family. Biochemical characterization, binding interactions, and tissue localization. *J Biol Chem*, **282**, 11805–11816.

109. Timpl R, Sasaki T, Kostka G and Chu ML (2003). Fibulins: a versatile family of extracellular matrix proteins. *Nat Rev*, **4**, 479–489.

110. Wagenseil JE and Mecham RP (2009). Vascular extracellular matrix and arterial mechanics. *Physiol Rev*, **89**, 957–989.

111. Bressan GM, Daga-Gordini D, Colombatti A, Castellani I, Marigo V and Volpin D (1993). Emilin, a component of elastic fibers preferentially located at the elastin-microfibrils interface. *J Cell Biol*, **121**, 201–212.

112. Zanetti M, Braghetta P, Sabatelli P, Mura I, Doliana R, Colombatti A, Volpin D, Bonaldo P and Bressan GM (2004). Emilin-1 deficiency induces elastogenesis and vascular cell defects. *Mol Cell Biol*, **24**, 638–650.

113. Zacchigna L, Vecchione C, Notte A, Cordenonsi M, Dupont S, Maretto S, Cifelli G, Ferrari A, Maffei A, Fabbro C, Braghetta P, Marino G, Selvetella G, Aretini A, Colonnese C, Bettarini U, Russo G, Soligo S, Adorno M, Bonaldo P, Volpin D, Piccolo S, Lembo G and Bressan GM (2006). Emilin-1 links TGF-beta maturation to blood pressure homeostasis. *Cell*, **124**, 929–942.

114. Robb BW, Wachi H, Schaub T, Mecham RP and Davis EC (1999). Characterization of an in vitro model of elastic fiber assembly. *Mol Biol Cell*, **10**, 3595–3605.

115. Kozel BA, Rongish BJ, Czirok A, et al. (2006). Elastic fiber formation: a dynamic view of extracellular matrix assembly using timer reporters. *J Cell Phys*, **207**, 87–96.

116. Czirok A, Zach J, Kozel BA, Mecham RP, Davis EC and Rongish BJ (2006). Elastic fiber macro-assembly is a hierarchical, cell motion-mediated process. *J Cell Physiol*, **207**, 97–106.

117. Li DY, Faury G, Taylor DG, Davis EC, Boyle WA, Mecham RP, Stenzel P, Boak B and Keating MT (1998). Novel arterial pathology in mice and humans hemizygous for elastin. *J Clin Invest*, **102**, 1783–1787.

118. Chrzanowski P, Keller S, Cerreta J, et al. (1980). Elastin content of normal and emphysematous lung parenchyma. *Am J Med*, **69**, 351–359.

119. Shifren A and Mecham RP (2006). The stumbling block in lung repair of emphysema: elastic fiber assembly. *Proc Am Thorac Soc*, **3**, 428–433.

120. Zhang M, Pierce RA, Wachi H, Mecham RP and Parks WC (1999). An open reading frame element mediates posttranscriptional regulation of tropoelastin and responsiveness to transforming growth factor beta1. *Mol Cell Biol*, **19**(11), 7314–7326.

121. Durbeej M (2010). Laminins. *Cell Tissue Res*, **339**, 259–268.

122. Patarroyo M, Tryggvason K and Virtanen I (2002). Laminin isoforms in tumor invasion, angiogenesis and metastasis. *Semin Cancer Biol*, **12**, 197–207.

123. Colognato H and Yurchenco PD (2000). Form and function: thelaminin family of heterotrimers. *Dev Dyn*, **218**, 213–234.

124. Yurchenco PD, Tsilibary EC, Charonis AS and Furthmayr H (1985). Laminin polymerization in vitro. Evidence for a twostep assembly with domain specificity. *J Biol Chem*, **260**, 7636–7644.

125. Smyth N, Vatansever HS, Murray P, Meyer M, Frie C, Paulsson M and Edgar D (1999). Absence of basement membranes after targeting the LAMC1 gene results in embryonic lethality due to failure of endoderm differentiation. *J Cell Biol*, **144**, 151–160.

126. Özbek S, Balasubramanian PG, Chiquet-Ehrismann R, Tucker RP and Adams JC (2010). The evolution of extracellular matrix. *Mol Biol Cell*, **21**(24), 4300–4305.

127. Pöschl E, Mayer U, Stetefeld J, Baumgartner R, Holak TA, Huber R and Timpl R (1996). Site-directed mutagenesis and structural interpretation of the nidogen binding site of the laminin γ1 chain. *EMBO J*, **15**, 5154–5159.

128. Van Agtmael T and Bruckner-Tuderman L (2010). Basement membranes and human disease. *Cell Tissue Res*, **339**, 167–188.

129. Sasaki T, Fässler R and Hohenester E (2004). Laminin: the crux of basement membrane assembly. *J Cell Biol*, **164**, 959–963.

130. Miner JH (2008). Laminins and their roles in mammals. *Microsc Res Tech*, **71**, 349–356.

131. Nguyen NM and Senior RM (2006). Laminin isoforms and lung development: all isoforms are not equal. *Dev Biol*, **294**, 271–279.

132. Aumailley M, Bruckner-Tuderman L, Carter WG, Deutzmann R, Edgar D, Ekblom P, Engel J, Engvall E, Hohenester E, Jones JC, Kleinman HK, Marinkovich MP, Martin GR, Mayer U, Meneguzzi G, Miner JH, Miyazaki K, Patarroyo M, Paulsson M, Quaranta V, Sanes JR, Sasaki T, Sekiguchi K, Sorokin LM, Talts JF, Tryggvason K, Uitto J, Virtanen I, Mark K von der, Wewer UM, Yamada Y and Yurchenco PD (2005). A simplified laminin nomenclature. *Matrix Biol*, **24**, 326–332.

133. Miner JH, Li C, Mudd JL, Go G and Sutherland AE (2004). Compositional and structural requirements for laminin and basement membranes during mouse embryo implantation and gastrulation. *Development*, **131**, 2247–2256.

134. Frieser M, Nockel H, Pausch F, Roder C, Hahn A, Deutzmann R and Sorokin LM (1997). Cloning of the mouse laminin alpha 4 cDNA. Expression in a subset of endothelium. *Eur J Biochem*, **246**, 727–735.

135. Sorokin LM, Pausch F, Frieser M, Kroger S, Ohage E and Deutzmann R (1997). Developmental regulation of the laminin alpha5 chain suggests a role in epithelial and endothelial cell maturation. *Dev Biol*, **189**, 285–300.

136. Hynes RO (1990). *Fibronectins*. Springer, Berlin Heidelberg New York.

137. Pankov R and Yamada KM (2002). Fibronectin at a glance. *J Cell Sci*, **115**, 3861–3863.

138. George EL, Georges-Labouesse EN, Patel-King RS, Rayburn H and Hynes RO (1993). Defects in mesoderm, neural tube and vascular development in mouse embryos lacking fibronectin. *Development*, **119**, 1079–1091.

139. Johansson S, Svineng G, Wennerberg K, Armulik A and Lohikangas L (1997). Fibronectin-integrin interactions. *Front Biosci*, **2**, d126–146.

140. Kosmehl H, Berndt A and Katenkamp D (1996). Molecular variants of fibronectin and laminin: structure, physiological occurrence and histopathological aspects. *Virchows Arch*, **429**, 311–322.

141. Dzamba BJ and Peters DM (1991). Arrangement of cellular fibronectin in Now 131 noncollagenous fibrils in human fibrohlast cultures. *J Cell Sci*, **100**, 605–612.

142. Wierzbicka-Patynowski I and Schwarzbauer JE (2003). The ins and outs of fibronectin matrix assembly. *J Cell Sci*, **116**, 3269–3276.

143. Geiger B, Bershadsky A, Pankov R and Yamada KM (2001). Transmembrane extracellular matrix-cytoskeleton crosstalk. *Nat Rev Mol Cell Biol*, **2**, 793–805.

144. Mao Y and Schwarzbauer JE (2005). Fibronectin fibrillogenesis, a cell-mediated matrix assembly process. *Matrix Biol*, **24**(6), 389–399.

145. Schwarzbauer JE (1991). Identification of the fibronectin sequences required for assembly of a fibrillar matrix. *J Cell Biol*, **113**, 1463–1473.

146. Sechler JL, Takada Y and Schwarzbauer JE (1996). Altered rate of fibronectin matrix assembly by deletion of the first type III repeats. *J Cell Biol*, **134**, 573–583.

147. Johansson S (1985). Demonstration of high affinity fibronectin receptors on rat hepatocytes in suspension. *J Biol Chem*, **260**, 1557–1561.

148. Singh P, Carraher C and Schwarzbauer JE (2010). Assembly of fibronectin extracellular matrix. *Annu Rev Cell Dev Biol*, **26**, 397–419.

149. Plow EF, Haas TA, Zhang L, Loftus J and Smith JW (2000). Ligand binding to integrins. *J Biol Chem*, **275**, 21785–21788.

150. Fogerty FJ, Akiyama SK, Yamada KM and Mosher DF (1990). Inhibition of binding of fibronectin to matrix assembly sites by anti-integrin (a5b1) antibodies. *J Cell Biol*, **111**, 699–708.

151. Wu C, Keivens VM, O'Toole TE, McDonald JA and Ginsberg MH (1995). Integrin activation and cytoskeletal interaction are essential for the assembly of a fibronectin matrix. *Cell*, **83**, 715–724.

152. Sage EH and Bornstein P (1991). Minireview: extracellular proteins that modulate cell-matrix interactions: SPARC, tenascin, and thrombospondin. *J Biol Chem*, **266**, 14831–14834.

153. Bornstein P (2009). Matricellular proteins: an overview. *J Cell Commun Signal*, **3**, 163–165.

154. Murphy-Ullrich JE (2001). The deadhesive activity of matricellular proteins: is intermediate cell adhesion an adaptive state? *J Clin Invest*, **10**, 785–790.

155. Liu A, Mosher DF, Murphy-Ullrich JE and Goldblum SE (2009). The counteradhesive proteins, thrombospondin 1 and SPARC/osteonectin, open the tyrosine phosphorylation-responsive paracellular pathway in pulmonary vascular endothelia. *Microvasc Res*, **77**, 13–20.

156. Bradshaw AD and Sage EH (2001). SPARC, a matricellular protein that functions in cellular differentiation and tissue response to injury. *J Clin Invest*, **107**, 1049–1054.

157. Jones PL and Jones FS (2000). Tenascin-C in development and disease: gene regulation and cell function. *Matrix Biol*, **19**, 581–596.

158. Denhardt DT, Noda M, O'Regan AW, Pavlin D and Berman JS (2001). Osteopontin as a means to cope with environmental insults: regulation of inflammation, tissue remodeling, and cell survival. *J Clin Invest*, **107**, 1055–1061.

159. Bornstein P and Sage EH (2002). Matricellular proteins: extracellular modulators of cell function. *Curr Opin Cell Biol*, **14**, 608–616.

160. Bornstein P (1995). Diversity of function is inherent in matricellular proteins: an appraisal of thrombospondin 1. *J Cell Biol*, **130**, 503–506.

161. Adams JC (2001). Thrombospondins: multifunctional regulators of cell interactions. *Annu Rev Cell Dev Biol*, **17**, 25–51.

162. Alford AI and Hankenson KD (2006). Matricellular proteins: extracellular modulators of bone development, remodeling, and regeneration. *Bone*, **38**, 749–757.

163. Mo FE and Lau LF (2006). The matricellular protein CCN1 is essential for cardiac development. *Circ Res*, **99**, 961–969.

164. Maurer P, Mayer U, Bruch M, Jeno P, Mann K, Landwehr R, Engel J and Timpl R (1992). High-affinity and low-affinity calcium binding and stability of the multidomain extracellular 40-kDa basement membrane glycoprotein (BM-40/SPARC/osteonectin). *Eur J Biochem*, **205**, 233–240.

165. Romberg RW, Werness PG, Lollar P, Riggs BL and Mann KG (1985). Isolation and characterization of native adult osteonectin. *J Biol Chem*, **260**, 2728–2736.

166. Stenner DD, Romberg RW, Tracy RP, Katzmann JA, Riggs BL and Mann KG (1984). Monoclonal antibodies to native noncollagenous bone-specific proteins. *Proc Natl Acad Sci USA*, **81**, 2868–2872.

167. Malaval L, Darbouret B, Preaudat C, Jolu JP and Delmas PD (1991). Intertissular variations in osteonectin: a monoclonal antibody directed to bone osteonectin shows reduced affinity for platelet osteonectin. *J Bone Miner Res*, **6**, 315–323.

168. Aeschlimann D, Wetterwald A, Fleisch H and Paulsson M (1993). Expression of tissue transglutaminase in skeletal tissues correlates with events of terminal differentiation of chondrocytes. *J Cell Biol*, **120**, 1461–1470.

169. Aeschlimann D, Kaupp O and Paulsson M (1995). Transglutaminase catalyzed matrix cross-linking in differentiating cartilage: identification of osteonectin as a major glutaminyl substrate. *J Cell Biol*, **129**, 881–892.

170. Mann K, Deutzmann R, Paulsson M and Timpl R (1987). Solubilization of protein BM-40 from a basement membrane tumor with chelating agents and evidence for its identity with osteonectin and SPARC. *FEBS Lett*, **218**(1), 167–172.

171. Maillard C, Malaval L and Delmas PD (1992). Immunological screening of SPARC/osteonectin in nonmineralized tissues. *Bone*, **13**, 257–264.

172. Schellings MW, Pinto YM and Heymans S (2004). Matricellular proteins in the heart: possible role during stress and remodeling. *Cardiovasc Res*, **64**, 24–31.

173. Murphy-Ullrich JE (2001). The de-adhesive activity of matricellular proteins: is intermediate cell adhesion an adaptive state? *J Clin Invest*, **107**(7), 785–790.

174. Palacek SP, Loftus JC, Ginsberg MH, Lauffenburger DA and Horwitz AF (1997). Integrin-ligand binding properties govern cell migration speed through cell-substratum adhesiveness. *Nature*, **385**, 537–540.

175. Murphy-Ullrich JE, Lane TF, Pallero MA and Sage EH (1995). SPARC mediates focal adhesion disassembly in endothelial cells through a follistatin-like region and the calcium-binding EF-hand. *J Cell Biochem*, **57**, 341–350.

176. Murphy-Ullrich JE, Gurusiddappa S, Frazier WA and Höök M (1993). Heparin-binding peptides from thrombospondin-1 and 2 contain focal adhesion-lablizing activity. *J Biol Chem*, **268**, 26784–26789.

177. Bellon G, Martiny L and Robinet A (2004). Matrix metalloproteinases and matrikines in angiogenesis. *Crit Rev Oncol Hematol*, **49**, 203–220.

178. Maquart FX, Pasco S, Ramont L, Hornebeck W and Monboisse JC (2004). An introduction to matrikines: extracellular matrix-derived peptides which regulate cell activity. Implication in tumor invasion. *Crit Rev Oncol Hematol*, **49**, 199–202.

179. Tran KT, Lamb P and Deng JS (2005). Matrikines and matricryptins: implications for cutaneous cancers and skin repair. *J Dermatol Sci*, **40**, 11–20.

180. Panayotou G, End P, Aumailley M, Timpl R and Engel J (1989). Domains of laminin with growth-factor activity. *Cell*, **56**, 93–101.

181. Giannelli G, Falk-Marzillier J, Schiraldi O, Stetler-Stevenson WG and Quaranta V (1997). Induction of cell migration by matrix metalloprotease-2 cleavage of laminin-5. *Science*, **277**, 225–228.

182. Tran KT, Lamb P and Deng J-S (2005). Matrikines and matricryptins: implications for cutaneous cancers and skin repair. *J Dermatol Sci*, **40**(1), 11–20.

183. Ricard-Blum S and Ballut L (2011). Matricryptins derived from collagens and proteoglycans. *Front Biosci*, **16**, 674–697.

184. Eckes B, Nischt R and Krieg T (2010). Cell-matrix interactions in dermal repair and scarring. *Fibrogenesis Tissue Repair*, **3**, 4.

185. Iozzo RV and Murdoch AD (1996). Proteoglycans of the extracellular environment: clues from the gene and protein side offer novel perspectives in molecular diversity and function. *FASEB J*, **10**, 598–614.

186. Oldberg A, Antonsson P, Hedbom E and Heinegard D (1990). Structure and function of extracellular matrix proteoglycans. *Biochem Soc Trans*, **18**, 789–792.

187. Schaefer L and Schaefer RM (2010). Proteoglycans: from structural compounds to signaling molecules. *Cell Tissue Res*, **339**, 237–246.

188. Heinegard D and Oldberg A (1993). Glycosylated matrix proteins. In: Royce PM and Steinmann B, eds. *Connective Tissue and Its Heritable Disorders: Molecular, Genetic, and Medical Aspects.* New York, NY: Wiley-Liss, pp. 189–209.

189. Hardingham TE and Fosang AJ (1992). Proteoglycans: many forms and many functions. *FASEB J*, **6**, 861–870.

190. Bidanset DJ, Guidry C, Rosenberg LC, Choi HU, Timpl R and Höök M (1992). Binding of the proteoglycan decorin to collagen type VI. *J Biol Chem*, **267**, 5250–5256.

191. Winnemoller M, Schmidt G and Kresse H (1991). Influence of decorin on fibroblast adhesion to fibronectin. *Eur J Cell Biol*, **54**, 10–17.

192. Winnemoller M, Schon P, Visher P and Kresse H (1992). Interactions between thrombospondin and the small proteoglycan decorin: interference with cell attachment. *Eur J Cell Biol*, **59**, 47–55.

193. Pogány G, Hernandez DJ and Vogel KG (1994). The in vitro interaction of proteoglycans with type I collagen is modulated by phosphate. *Arch Biochem Biophys*, **313**, 101–111.

194. Järveläinen H, Puolakkainen P, Pakkanen S, Brown EL, Höök M, Iozzo RV, Sage EH and Wight TN (2006). A role for decorin in cutaneous

wound healing and angiogenesis. *Wound Repair Regen*, **14**, 443–452.

195. Schaefer L and Iozzo RV (2008). Biological functions of the small leucine-rich proteoglycans: from genetics to signal transduction. *J Biol Chem*, **283**, 21305–21309.

196. Bulow HE and Hobert O (2006). The molecular diversity of glycosaminoglycans shapes animal development. *Annu Rev Cell Dev Biol*, **22**, 375–407.

197. Aumailley M and Gayraud B (1998). Structure and biological activity of the extracellular matrix. *J Mol Med*, **76**, 253–265.

198. Scott JE (1996). Proteodermatan and proteokeratan sulfate (decorin, lumican/fibromodulin) proteins are horseshoe shaped. Implications for their interactions with collagen. *Biochemistry*, **35**, 8795–8799.

199. McEwan PA, Scott PG, Bishop PN and Bella J (2006). Structural correlations in the family of small leucine-rich repeat proteins and proteoglycans. *J Struct Biol*, **155**(2), 294–305.

200. Iozzo RV (1999). The biology of the small leucine-rich proteoglycans. *J Biol Chem*, **274**(27), 18843–18846.

201. Kadler KE, Hill A and Canty-Laird EG (2008). Collagen fibrillogenesis: fibronectin, integrins, and minor collagens as organizers and nucleators. *Curr Opin Cell Biol*, **20**, 495–501.

202. Chakravarti S, Zhang G, Chervoneva I, Roberts L and Birk DE (2006). Collagen fibril assembly during postnatal development and dysfunctional regulation in the lumican-deficient murine cornea. *Dev Dyn*, **235**, 2493–2506.

203. Iozzo RV (1998). Matrix proteoglycans: from molecular design to cellular function. *Annu Rev Biochem*, **67**, 609–652.

204. Wu RR and Couchman JR (1997). cDNA cloning of the basement membrane chondroitin sulfate proteoglycan core protein, bamacan: a five domain structure including coiled-coil motifs. *J Cell Biol*, **136**, 433–444.

205. Wu YJ, La Pierre DP, Wu J, Yee AJ and Yang BB (2005). The interaction of versican with its binding partners. *Cell Res*, **15**, 483–494.

206. Bernfield M, Gotte M, Park PW, Reizes O, Fitzgerald ML, Lincecum J and Zako M (1999). Functions of cell surface heparan sulfate proteoglycans. *Annu Rev Biochem*, **68**, 729–777.

207. Alexopoulou AN, Multhaupt HA and Couchman JR (2007). Syndecans in wound healing, inflammation and vascular biology. *Int J Biochem Cell Biol*, **39**, 505–528.

208. Xian X, Gopal S and Couchman JR (2010). Syndecans as receptors and organizers of the extracellular matrix. *Cell Tissue Res*, **339**, 31–46.

209. Couchman JR (2003). Syndecans: proteoglycan regulators of cell surface microdomains? *Nat Rev Mol Cell Biol*, **4**, 926–937.

210. Filmus J, Capurro M and Rast J (2008). Glypicans. *Genome Biol*, **9**, 224; Rodgers KD, San Antonio JD and Jacenko O (2008). Heparan sulfate proteoglycans: a GAGgle of skeletal-hematopoietic regulators. *Dev Dyn*, **237**, 2622–2642.

211. Ornitz DM (2000). FGFs, heparan sulfate and FGFRs: complex interactions essential for development. *Bioessays*, **22**, 108–112.

212. Jia J, Maccarana M, Zhang X, Bespalov M, Lindahl U and Li JP (2009). Lack of L-iduronic acid in heparan sulfate affects interaction with growth factors and cell signaling. *J Biol Chem*, **284**, 15942–15950.

213. Sugahara K and Mikami T (2007). Chondroitin/dermatan sulfate in the central nervous system. *Curr Opin Struct Biol*, **17**, 536–545.

214. Kwok JC, Afshari F, Garcia-Alias G and Fawcett JW (2008). Proteoglycans in the central nervous system: plasticity, regeneration and their stimulation with chondroitinase ABC. *Restor Neurol Neurosci*, **26**, 131–145.

215. Bandtlow CE and Zimmermann DR (2000). Proteoglycans in the developing brain: new conceptual insights for old proteins. *Physiol Rev*, **80**, 1267–1290.

216. Roughley PJ (2006). The structure and function of cartilage proteoglycans. *Eur Cell Mater*, **12**, 92–101.

217. van den Heuvel LP, Veerkamp JH, Monnens LA and Schroder CH (1988). HSPG from human and equine glomeruli and tubules. *Int J Biochem*, **20**, 527–532.

218. Kallunki P and Tryggvason K (1992). Human basement membrane HSPG core protein containing multiple domains resembling elements of the low density lipoprotein receptor, laminin, neural cell adhesion molecules and epidermal growth factor. *J Cell Biol,* **116**, 552–271.

219. Murdoch AD, Liu B, Schwarting R, Tuan, RS and Iozzo RV (1994). Widespread expression of perlecan proteoglycan in basement membranes and extracellular matrices of human tissues as detected by a novel monoclonal antibody against domain III and by in situ hybridization. *J Histochem Cytochem*, **42**, 239–249.

220. Yurchenco PD, Cheg YS and Ruben GC (1987). Self assembly of a high molecular weight basement membrane HSPG into dimmers and oligomers. *J Biol Chem*, **262**, 5036–5043.

221. Melrose J, Roughley P, Knox S, Smith S, Lord M and Whitelock J (2006). The structure, location, and function of perlecan, a prominent pericellular proteoglycan of fetal, postnatal, and mature hyaline cartilages. *J Biol Chem*, **281**, 36905–36914.

222. Smith SM, West LA, Govindraj P, Zhang X, Ornitz DM and Hassell JR (2007a). Heparan and chondroitin sulfate on growth plate perlecan mediate binding and delivery of FGF-2 to FGF receptors. *Matrix Biol*, **26**, 175–184.

223. Tholozan FM, Gribbon C, Li Z, Goldberg MW, Prescott AR, McKie N and Quinlan RA (2007). FGF-2 release from the lens capsule by MMP-2 maintains lens epithelial cell viability. *Mol Biol Cell*, **18**, 4222–4231.

224. Smith SM, West LA and Hassell JR (2007b). The core protein of growth plate perlecan binds FGF-18 and alters its mitogenic effect on chondrocytes. *Arch Biochem Biophys*, **468**, 244–251.

225. Patel VN, Knox SM, Likar KM, Lathrop CA, Hossain R, Eftekhari S, Whitelock JM, Elkin M, Vlodavsky I and Hoffman MP (2007). Heparanase cleavage of perlecan heparan sulfate modulates FGF10 activity during ex vivo submandibular gland branching morphogenesis. *Development*, **134**, 4177–4186.

226. Ruegg MA and Bixby JL (1998). Agrin orchestrates synaptic differentiation at the vertebrate neuromuscular junction. *Trends Neurosci*, **21**, 22–27.

227. Barber AJ and Lieth E (1997). Agrin accumulates in the brain microvascular basal lamina during development of the blood-brain barrier. *Dev Dyn*, **208**, 62–74.

228. Groffen AJ, Buskens CA, van Kuppevelt TH, Veerkamp H, Monnens LA and van den Heuvel LP (1998). Primary structure and high expression of human agrin in basement membranes of adult lung and kidney. *Eur J Biochem*, **254**, 123–128.

229. Groffen AJ, Ruegg MA, Dijkman H, et al. (1998). Agrin is a major HSPG in the human GBM. *J Histochem Cytochem*, **46**, 19–27.

230. Durbeej M, Henry MD, Ferletta M, Campbell KP and Ekblom P (1998). Distribution of dystroglycan in normal adult mouse tissues. *J Histoche Cytochem*, **46**, 449–457.

231. Kanwar YS, Linker A and Farquhar MG (1980). Increased permeability of the GBM to ferritin after removal of glycosaminoglycans (HS) by enzyme digestion. *J Cell Biol*, **86**, 688–693.

232. Hassell JR, Robey PG, Barrach HJ, Wilczek J, Rennard SI and Martin GR (1980). Isolation of a HS containing proteoglycan from basement membrane. *Proc Natl Acad Sci USA*, **77**, 4494–4498.

233. Perrimon N and Bernfield M (2001). Cellular functions of proteoglycans: an overview. *Semin Cell Dev Biol*, **12**, 65–67.

234. Kjellen L and Lindahl U (1991). Proteoglycans: structures and interactions. *Annu Rev Biochem*, **60**, 443–475.

235. Gandhi NS and Mancera RL (2008). The structure of glycosaminoglycans and their interactions with proteins. *Chem Biol Drug Des*, **72**, 455–482.

236. Lau KS, Partridge EA, Grigorian A, Silvescu CI, Reinhold VN, Demetriou M and Dennis JW (2007). Complex N-glycan number and degree of branching cooperate to regulate cell proliferation and differentiation. *Cell*, **129**, 123–134.

237. Flint MH, Gillard GC and Merrilees MJ (1980). Effects of local environmental factors on connective tissue organisation and glycoaminoglycan synthesis. In: Parry DAD and Creamer LK, eds. *Fibrous Proteins: Scientific, Industrial, and Medical Aspects*. London, England: Academic Press, pp. 107–119.

238. McDonald J and Hascall VC (2002). Hyaluronan minireview series. *J Biol Chem*, **277**, 4575–4579.

239. Fraser JRE and Laurent TC (1996). Hyaluronan. In: Comper WD, ed. *Extracellular Matrix*, Vol. 1: Tissue Function. Amsterdam, the Netherlands: Harwood Academic Publishers.

240. Necas J, Bartosikova L, Brauner P, and Kolar J (2008). Hyaluronic acid (hyaluronan): a review. *Veterinarni Medicina*, **53**(8), 397–411.

241. Hynes RO and Zhao Q (2000). The evolution of cell adhesion. *J Cell Biol*, **150**, F89–96.

242. Roy P, et al. (1999). Effect of cell migration on the maintenance of tension on a collagen matrix. *Ann Biomed Eng*, **27**, 721–730.

243. Raucher D and Sheetz MP (2000). Cell spreading and lamellipodial extension rate is regulated by membrane tension. *J Cell Biol*, **148**, 127–136.

244. Rodan GA and Martin TJ (2000). Therapeutic approaches to bone diseases. *Science*, **289**, 1508–1514.

245. Marchisio PC, et al. (1984). Cell-substratum interaction of cultured avian osteoclasts is mediated by specific adhesion structures. *J Cell Biol*, **99**, 1696–1705.

246. Teitelbaum SL (2000). Bone resorption by osteoclasts. *Science*, **289**, 1504–1508.

247. Warren CM and Iruela-Arispe ML (2014). Podosome rosettes precede vascular sprouts in tumour angiogenesis. *Nat Cell Biol*, **16**, 928–930.

248. Kefalides NA and Borel JP (2005). Contacts of basement membrane molecules with cell membranes. *Curr Top Membr*, **56**, 287–319.

249. Kleyman I and Brannagan TH (2014). In *Encyclopedia of the Neurological Sciences*, 2nd edn., pp. 1054–1056.

250. Adams JC (2002). Molecular organisation of cell–matrix contacts: essential multiprotein assemblies in cell and tissue function. *Expert Rev Mol Med*, **4**, 1–24.

251. Hynes RO (1990). *Fibronectins*, Springer-Verlag, Berlin.

252. Inki P and Jalkanen M (1996). The role of syndecan-1 in malignancies. *Ann Med*, **28**, 63–67.

253. Sanders RJ, Mainiero F and Giancotti FG (1998). The role of integrins in tumorigenesis and metastasis. *Cancer Invest*, **16**, 329–344.

254. Ruoslahti E (1999). Fibronectin and its integrin receptors in cancer. *Adv Cancer Res*, **76**, 1–20.

255. Burridge K and Chrzanowska-Wodnicka M (1996). Focal adhesions, contractility, and signaling. *Annu Rev Cell Dev Biol*, **12**, 463–518.

256. Pelham RJ Jr and Wang YL (1998). Cell locomotion and focal adhesions are regulated by the mechanical properties of the substrate. *Biol Bull*, **194**, 348–349; discussion 349–350.

257. Ingber E, Dike L, Hansen L, Karp S, Liley H, Maniotis A, McNamee H, Mooney D, Plopper G, Sims J and Wang N (1994). Cellular tensegrity: exploring how mechanical changes in the cytoskeleton regulate cell growth, migration, and tissue pattern during morphogenesis. *Int Rev Cytol*, **150**, 173–224.

258. Ingber DE (2003). Tensegrity I. Cell structure and hierarchical systems biology. *J Cell Sci*, **116**, 1157–1173.

259. Sultan C, Stamenović D and Ingber DE (2004). A computational tensegrity model predicts dynamic rheological behaviors in living cells. *Ann Biomed Eng*, **32**(4), 520–530.

260. Alberts B, Johnson A, Lewis J, et al. (2002). *Molecular Biology of the Cell*, 4th edn., New York: Garland Science.

261. Zamir E and Geiger B (2001). Components of cell-matrix adhesions. *J Cell Sci*, **114**, 3577–3579.

262. Lu C, Takagi J and Springer TA (2001). Association of the membrane proximal regions of the alpha and beta subunit cytoplasmic domains constrains an integrin in the inactive state. *J Biol Chem*, **276**, 14642–14648.

263. Takagi J, Erickson HP and Springer TA (2001). C-terminal opening mimics 'inside-out' activation of integrin alpha5beta1. *Nat Struct Biol*, **8**, 412–416.

264. Danen EHJ (2013). Integrins: an overview of structural and functional aspects. In *Madame Curie Bioscience Database [Internet]*. Austin (TX): Landes Bioscience; 2000–2013.

265. Clark EA and Brugge JS (1995). Integrins and signal transduction pathways: the road taken. *Science*, **268**, 233–239.

266. Fuchs E, et al. (1997). Integrators of epidermal growth and differentiation: distinct functions for beta 1 and beta 4 integrins. *Curr Opin Genet Dev*, **7**, 672–682.

267. Borradori L and Sonnenberg A (1999). Structure and function of hemidesmosomes: more than simple adhesion complexes. *J Invest Dermatol*, **112**, 411–418.

268. Gullberg D, Gehlsen KR, Turner DC, Ahlen K, Zijenah LS, Barnes MJ and Rubin K (1992). Analysis of $\alpha 1\beta 1$, $\alpha 2\beta 1$ and $\alpha 3\beta 1$ integrins in cell–collagen interactions: identification of conformation dependent $\alpha 1\beta 1$ binding sites in collagen type I. *EMBO J*, **11**, 3865–3873.

269. Eble JA, Golbik R, Mann K and Kühn K (1993). The $\alpha 1\beta 1$ integrin recognition site of the basement membrane collagen molecule [$\alpha 1$(IV)]$2\alpha 2$(IV). *EMBO J*, **12**, 4795–4802.

270. Ruggiero F, Champliaud MF, Garrone R and Aumailley M (1994). Interactions between cells and collagen V molecules or single chains involve distinct mechanisms. *Exp Cell Res*, **210**, 215–223.

271. Vandenberg P, Kern A, Ries A, Luckenbill-Edds L, Mann K and Kühn K (1991). Characterization of a type IV collagen major cell binding site with affinity to the $\alpha 1\beta 1$ and the $\alpha 2\beta 1$ integrins. *J Cell Biol*, **113**, 1475–1483.

272. Pfaff M, Aumailley M, Specks U, Knolle J, Zerwes HG and Timpl R (1993). Integrin and Arg-Gly-Asp dependence of cell adhesion to the native and unfolded triple helix of collagen type VI. *Exp Cell Res*, **206**, 167–176.

273. Dickinson CD, Veerapandia B, Dai XP, Hamlin RC, Xuong NH, Ruoslahti E and Ely KR (1994). Crystal structure of the tenth type III cell adhesion module of human fibronectin. *J Mol Biol*, **236**, 1079–1092.

274. Main AL, Harvey TS, Baron M, Boyd J and Campbell ID (1992). The three-dimensional structure of the tenth type III module of fibronectin: an insight into RGD-mediated interactions. *Cell*, **71**, 671–678.

275. Leahy DJ, Hendrickson WA, Aukhil I and Erickson HP (1992). Structure of a fibronectin type III domain from tenascin phased by MAD analysis of the selenomethionyl protein. *Science*, **258**, 987–991.

276. Nesbitt S, et al. (1993). Biochemical characterization of human osteoclast integrins. Osteoclasts express alpha v beta 3, alpha 2 beta 1, and alpha v beta 1 integrins. *J Biol Chem*, **268**, 16737–16745.

277. Bateman JF, Boot-Handford RP and Lamandé SR (2009). Genetic diseases of connective tissues: cellular and extracellular effects of ECM mutations. *Nat Rev Genet*, **10**, 173–183.

278. Hirako Y, Usukura J, Nishizawa Y and Owaribe K (1996). Demonstration of the molecular shape of BP180, a 180-kDa bullous pemphigoid antigen and its potential trimer formation. *J Biol Chem*, **271**, 13739–13745.

279. Li K, Tamai K, Tan EM and Uitto J (1993). Cloning of type XVII collagen. Complementary and genomic DNA sequences of mouse 180-kilodalton bullous pemphigoid antigen (BPAG2) predict an interrupted collagenous domain, a transmembrane segment, and unusual features in the 5′-end of the gene and the 3′-untranslated region of the mRNA. *J Biol Chem*, **268**, 8825–8834.

280. Carter WG, Kaur P, Gil SG, Gahr PJ and Wayner EA (1990). Distinct functions for integrins $\alpha 3\beta 1$ in focal adhesions and $\alpha 6\beta 4$/bullous pemphigoid antigen in a new stable anchoring contact (SAC) of keratinocytes: relation to hemidesmosomes. *J Cell Biol*, **111**, 3141–3154.

Chapter 3

ECM Mimicking for 3D Cell Culture

3.1 3D Cell Culture (Introduction)

Conventional cell culture with only 2D space provides unnatural conditions for growing cells, which most often leads to physiologically compromised cells [1]. The importance of 3D space and the microenvironment was realized and acknowledged only over the last two decades when observations like the increase in collagen production by vascular smooth muscle cells (SMCs) on seeding within the scaffold of PEG-based hydrogels were recorded [2]. That growth specificity could be achieved through ECM management was also demonstrated through experiments where adhesion and growth of vascular SMCs but not fibroblasts or platelets [3] was observed in hydrogels modified with elastin-derived peptide. Such studies drew attention to the shortcomings of growing cells on flat surfaces and the likelihood of missing important information regarding cell physiology and its natural response. It also acknowledged the need to develop 3D matrices and scaffolds that can help in studying the cells in a native-like microenvironment. An appropriately constituted ECM mimicking 3D cell culture technology, which can allow customization for the growth of specified cells, can potentially revolutionize our understanding about the cell and its behavior vis à vis its microenvironment [4].

3D Cell Culture: Fundamentals and Applications in Tissue Engineering and Regenerative Medicine
Ranjna C. Dutta and Aroop K. Dutta
Copyright © 2018 Pan Stanford Publishing Pte. Ltd.
ISBN 978-981-4774-53-6 (Hardcover), 978-1-315-14682-9 (eBook)
www.panstanford.com

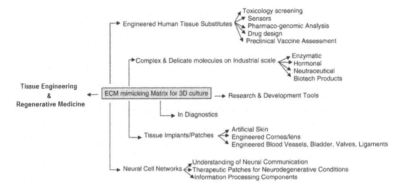

Figure 3.1 Potential applications of ECM-mimicking 3D matrix (from Dutta & Dutta, *Biotechnol Adv*, 2009, **27**, 334–339).

Many different materials, both natural and synthetic, and processes are being explored all over the world since the new quest. One of the important facts gleaned from the journey is that the scaffold biomaterial needs to possess holding strength while remaining cell interactive. A technology that can be as easy to use as the conventional cell culture plates but provide information as reliable as one could get from 3D culture is the need. With reproducible scaffold composition and surface topography, an ECM mimicking matrix that qualifies in the necessities of strength, porosity, stiffness, and degradation dynamics would find applications in many diverse areas (Fig. 3.1).

3.1.1 Significance of ECM Mimicking

Till now the engagement of cells with the immediate extracellular matrix (ECM) in 3D space and also with the neighboring cells has been overlooked inadvertently while evaluating their response in the presence of new drug entities or toxins or intrinsic physiological entities like enzymes and hormones. Nevertheless, the absence of an appropriate ECM milieu is expected to have little impact on genetic and intracellular molecular biology studies preventing us from getting fundamentally wrong.

The significance of ECM molecules and their spatial arrangement in ex vivo cell culture was realized for the first time when cells were grown in the presence of one or more ECM components. Fibroblasts produced more proteins when cultured on collagen fibrils than on a plastic surface [5]. It is further observed that cells cultured in (or on) thick and malleable matrices become differentiated, whereas cells grown on the ECM delivered as a thin, planar coating on plastic do not [6]. For example, only those mammary cells which were embedded in a laminin rich basement membrane are observed to adopt a spherical, polarized structure that resembles the normal mammary alveolus (or acinus) and are also capable of mammary-gland-specific functions (e.g., producing milk in response to lactogenic hormones) [7]. Furthermore, the phenotype of the newly generated cell is found to be extremely sensitive to tissue level elasticity [8]. All these studies underlined the importance of the microenvironment that is present in the 3D space surrounding the growing cells. These examples also pointed out an alarming/serious limitation associated with the cell culture done by conventional 2D methods, i.e., in isolation using a liquid medium. They also suggested that appropriate culture conditions are capable of facilitating the expression of more genes and also that the adeqautely tuned microenvironment is the key in the development of physiologically active artificial tissue. Tissue specific ECM can sustain the functional cells and apparently has an ability to reorganize the inoculated cells into a fully functional tissue. Regeneration of a beating heart from a carefully de-cellularized heart-ECM supports this perception [9]. Such studies conclusively establish the importance of the ECM for healthy functioning of a tissue and suggest that tissue specific ECM mimicking holds great promise in recreating functional tissues ex vivo.

Considering the role and importance of ECM, elaborated in Chapter 1, there is a clear need for ECM mimicking for in vitro cell culture. Culturing cells in 3 dimensional (3D) environments could provide a more conducive native-like setting to the cells. It is well accepted now that a near physiological response could be expected from cells growing in 3D as compared to a flat surface where they get the freedom to expand only in 2 dimensions. That 3D matrices or scaffolds representing physical mimicking of the ECM are better than 2D is also proved by using polycaprolactone (PCL) [10] and poly(lactic-*co*-glycolic acid) (10:90)–poly(ε-caprolactone) (PLGA–PCL) scaffolds

[11]. However, for engineering a physiologically functional tissue in vitro, the physical modeling of the extracellular matrix (ECM) is not sufficient and would require integrated biochemical and mechanical cues as well. As discussed in the previous chapter, ECM is a complex, fibrous mesh-like network, which is required to keep the cells stretched in defined directions and acquire a specific shape. In layman's language it can be stated that ECM helps to keep the cells alert to remain responsive to the environment. It also mediates and dictates their fate through a range of functional interactions. It plays a crucial role in regulating almost all aspects of cell response including growth, differentiation and tissue development. The ECM network is now established as the most influential component of the cellular microenvironment. Mutations and faulty mechanisms at any level from transcription to expresion of the ECM constituents are also implicated in a variety of pathologies.

The ECM of various tissues and organs differ not only in their composition but quantity as well. ECM rich tissues like bone, cartilage and cornea need a dense scaffold as their mimic. Physiologically functional tissues, which are cell dominant on the other hand, need crucial ECM components, preferably arranged in the pericellular region. The glue-like ECM in these tissues needs to support and bring co-ordination among the affiliated cells as its primary function. Scaffolds for their ex vivo reconstruction should therefore favor intercellular connections while permitting easy remodeling so that cells can rearrange in a tissue-like dense and compact format [12]. Independent studies with scaffolds of varying biomaterials have highlighted the importance of mechanical and geometrical cues for cell adhesion and proliferation. Cells are mechano-sensitive and respond to force-induced changes in protein conformation and geometry-dependent interactions around their local vicinity [13]. In multicellular organisms, cell-ECM and cell-cell interactions establish the polarity and cells cultured in conditions that do not facilitate these interactions normally remain unpolarized and die by anoikis or transform into a tumor [14].

Thus 3D cell culture in an ECM mimicking environment is at the helm of creating artificial tissue and can prove the ultimate tool in regenerative medicine and tissue engineering. Ex vivo creation of natural tissue alike is possible only when cells are allowed to grow in a biomimicking extracellular environment.

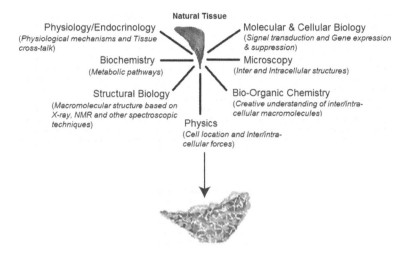

Figure 3.2 Interdisciplinary knowledge helps in mimicking ECM to engineer complex tissues.

Clues for physical, biochemical and mechanical mimicking of ECM can actually be derived from interdisciplinary knowledge and understanding (Fig. 3.2). The imaging and microscopy helps in identification of the cellular arrangement and their alignment vis-a-vis ECM while molecular and cell biology provides an insight into their dynamic interaction at the molecular level. Physiological and biochemical study provides information about the functional and metabolic aspects of cells and tissue while spectral analysis including NMR, X-ray, IR and other structural studies help in deciphering molecular interactions.

3.1.2 ECM Mimicking/Reconstitution

Culturing cells that differentiate and maintain in vivo like behavior is thus not possible by conventional flat surface methods and require an ECM mimicking microenvironment in 3D space. With limited understanding of molecular complexity and tissue specificity, replicating or reconstituting the ECM for artificial tissue construction appears to be a far-fetched goal. Nevertheless, physical, biochemical and/or mechanical mimicking of the ECM can be contemplated based on previous learnings. Knowledge acquired under different topics and sub-topics need to be effectively integrated for successful tissue engineering (Fig. 3.2). The existing understanding, however, permits only partial and qualitative duplication of the ECM which could be achieved by (1) adding speculative amounts of known ECM components to culture media; (2) using media supplements known to stimulate cells for secreting ECM; (3) growing cells on pre-culture beds; (4) using tissue explants, e.g., matrigel or decellularized functional tissues and (5) creating a biomaterial scaffold tailored with natural ECM components. The first two approaches do not provide the semi-rigid structural availability of ECM as found in vivo whereas the third and fourth deal with undefined ECM components. It is in the final approach that a biomaterial can be integrated with appropriately chosen, well characterized ECM components. Efforts have been made to imitate at least some aspects of ECM; for example to provide a spatial third dimension to the growing cells, scaffolds of different architectures have been created. Natural and synthetic biopolymer scaffolds of porous and fibrous morphology are evaluated to support the cells ex vivo. Mostly architectural (physical) and in few cases functional (biochemical) or mechanical mimicking of ECM has been attempted so far. The recovery of native ECM through decellularization has also been achieved recently.

3.1.2.1 Physical mimicking

Cells in vivo are primarily supported by adjoining ECM from all directions. The intercellular ECM span and holds multiple cells together to define the shape of a tissue. The ECM that supports tissues and organs in the human body ranges from soft gel-like to hard-bony material in its texture. This wide range of texture also encompasses membranous, fibrous, pelleted and elastic arrangements of ECM biomaterial in different tissues. An unambiguous link between tissue and the material characteristics of its ECM has already been established, re-emphasizing the importance of its texture [15]. Engineering a tissue ex vivo would therefore need scaffolds or matrices that match the shape and morphology of the target tissue while allowing the cells to expand and grow in 3D. Accordingly, fibrous or porous architectures and also their combinations may be utilized.

Though the cells contributing to a tissue may differ morphologically, they still complement each other metabolically in a concerted manner for accomplishing a defined physiological function. Arranged on a macroscale, each such group of cells or tissue is surrounded by a semi-permeable membrane. This barrier not only regulates the two-way exchange of ions and bio-chemicals but also imparts them a unit identity and also the shape of an organ. Unlike skin and vasculature boundaries/walls (arteries and veins) where cells seem to expand longitudinally through ECM in two dimensions only, in the rest of the organs they clearly take a complex three-dimensional form. The freedom of elastic expansion or stretching of a tissue/organ is dictated by the characteristic arrangement of ECM-biopolymers like elastin and collagens. Microscopic examination of the skin also reveals a supporting 3D matrix at its base. The basement membrane lying below is found to be important in assisting the skin to grow and differentiate in a peculiar manner. The epidermal cells keep multiplying while pushing the topmost layers that gradually keratinize and cyclically shed off. This helps in the renewal and generation of new skin; the process, however, slows down with ageing. Thus, the third dimension and also time as the fourth dimension play an important role in the overall function of tissues and organs. Scaffolds, that can provide physical space in three dimensions is therefore the very least but critical requirement as it allows cell-cell proximity promoting self-assembly and in turn tissue function [16].

Hydrogels as physical mimics of ECM have been evaluated extensively for culturing cells. They can absorb water in excess while retaining porous architecture and hence proved good substrates for tissue engineering (TE) [17, 18]. They could also be delivered in a minimally invasive manner and therefore find use in drug/growth factor delivery and a variety of other applications [19].

Physical mimicking of ECM can also be achieved by creating porous or fibrous scaffolds by using different synthetic and natural materials. The third dimension, added even by simple blowing of polystyrene fibers, could improve the cell response in comparison to that obtained by growing them on conventional devices. It has been proved that cell morphology could be retained to a great extent simply by providing a third dimension. Such all-synthetic scaffold devices could reduce the performance gaps compared to their counterparts existing in vivo. 3D-Biotek is the company that provides such plastic devices manufactured exclusively from polystyrene, a polymer already in use for decades as 2D cell culture plates (www.3Dbiotek.com).

3.1.2.2 Biochemical mimicking

Physical mimicking in the form of scaffolds provides partial imitation of an in vivo like 3D environment. The scaffold created from synthetic polymers offers only a rigid and passive support with minimal cell-interactivity. Plastic scaffolds are rigid and cannot be hydrated. Hydration is essential for the flexibility required by growing cells. Though they provide 3D spaces for cells to expand, yet being non-interactive and resistant to remodeling, synthetic polymer–based scaffolds do not allow appropriate differentiation. Scaffolds created using natural polymers- like collagen and gelatin on the other hand are better mimics of in vivo ECM support. Being cell-interactive, they offer many advantages over all-synthetic scaffolds and prove superior in mimicking the natural ECM.

Cell interactivity in naturally derived materials is due to their flexible demure that helps and co-operates in the rearrangement and organization of growing cells. This understanding led era of cell-culture in the presence of one or more ECM components or scaffolds created thereof. Scaffolds from ECM derived biopolymers display additional biochemical impressions which may help in differentiation. The integrated gaps and pores of the scaffolds provide free space for the cells to grow, while the biodegradable matrix allows them to multiply and expand. ECM Analog^{TM} technology (www.excellmatrix.com) is the only available technology at present which offers both these features on one platform. Nevertheless, given the complexity and tissue specific variations in the constituting components of ECM, generating a physiologically functional tissue in vitro is still challenging.

3.1.2.3 Mechanical mimicking

It is known that different tissues have different elasticity which is destined by the molecular composition of ECM that is tissue specific. Matching the mechanical characteristic of tissue specific ECM is therefore another essential feature to be incorporated in the scaffold. This is significant especially when 3D cell culture is to be used for tissue-like growth ex vivo. Mechanical strength depends on the microstructure, porosity and the mechanical design parameters like elasticity (young modulus) and stiffness (stress-strain relationship) of the scaffold material.

The mechanical characteristics of ECM in different human tissues vary widely and are crucial for normal physiological function-ing. Though limited attention is paid to achieve the mechanical mimicking of the ECM, it is still important in the context of ex vivo tissue engineering. Tissue level elasticity in 3D scaffolds is reported to influence even the cell lineage specification. Soft matrices that mimic the brain are found to be neurogenic whereas stiffer matrices that mimic muscle are myogenic and comparatively rigid matrices that mimic bones are osteogenic [8, 20]. It is now comprehensible that the rigidity and suppleness of a scaffold designed for growing skin may not be optimum for growing other tissues, for example the heart or bone. The mechanical activity of a tissue depends on the relative composite behavior of cells and its surrounding macromolecular support matrix. However, as of now we do not have sufficient understanding of the mechanical behavior of different tissues, which is likely to impact their normal and healthy physiological functioning. Nevertheless, it is known that a fibroid or keloid-like growth have an adverse influence on the physiological functioning of the tissue. The detrimental effect of fat deposit on mechanics of tissue vasculature that leads to altered blood pressure is well known. This establishes that the 3D scaffold should be of adequate mechanical strength for engineering a particular tissue.

Engineered substitutes for hard tissue- like bones or implants for replacing or supporting damaged bones are effective and successful only if their mechanical strength matches with the native bone. Mechanical correspondence helps to integrate the synthetic/fabricated implant with the host, which is imperative for its success. The most common example that establishes the need of mechanical matching in scaffolds for 3D cell-culture and its grafts is the relative success of bone implants made of stainless steel, titanium and its alloys. However, recently, ceramics, the inorganic, non-metallic crystalline material with high strength has found use in creating the scaffold for handling bone defects. Porcelain and other ceramic materials are also found valuable in creating dental prostheses [21]. Calcium phosphate (CaP), tricalcium phosphate (β-TCP) and hydroxyapatite (HA) are some of the preferred ceramics for creating bone and cartilage substitutes. Besides being lighter and sturdier, the improved and appropriate mechanical matching of ceramics and metal-enforced ceramic implants is still the basis of success for newly developed implants. Though ceramics are brittle yet with a slow rate of degradation, they can persist in vivo for months to years [22–24]. Considering these issues, only nanoceramics which demonstrate enhanced osteoblast proliferation and long term functions in vitro are now recommended for bone implants [25]. A hybrid biomaterial like the collagen-hydroxyapatite scaffold has demonstrated improved healing of bone defects in comparison to collagen alone [26]. Our group is also working towards cell-sensitive nanoceramic blends, presuming that an appropriate mix of nanoceramic with cell-interactive biomaterial will impart the scaffold requisite in vivo like microenvironment and strength. A 3D conductive scaffold created by embedding poly(3,4-ethylenedioxythiophene) and poly(4-styrene sulfonate) (PEDOT:PSS), a conductive polymer in the optimized nanocomposite of gelatin and bioactive glass has recently been tested for large bone defects [27]. A scaffold produced from beta-tricalcium-phosphate (β-TCP), a biodegradable, synthetic calcium phosphate ceramic which is osteoconductive and alginate composites have also been tested for bone tissue engineering [28]. The mechanical property of scaffolds could be attributed to the material, which constitutes the pore walls and the edges and also to the relative density as well as any anisotropic nature of the material, if present. Gibson and Ashby have classified porous solids into two groups, honeycombs and foams [29]. Mechanical properties of the scaffold need to be adequate enough to account for the changes that may occur due to contractile forces exerted by seeded cells [30, 31].

An interesting study by Harris et al. showed that the monolayer elasticity is two-orders of magnitude larger than the elasticity of their isolated cellular components [32]. They demonstrated that the monolayers could withstand more than doubling in length before failing through the rupture of intercellular junctions and also that the monolayer stretching is mediated through keratin filaments. This means an appropriate alignment with well placed intercellular junctions is capable of imparting cumulative mechanical strength.

Efforts are on to control the mechanical properties of the 3D scaffold through various design principles [33, 34]. For example, in order to control shape-mechanics through differential swelling or shrinkage multiple polymer components with distinct compositions are integrated within the scaffold [35]. For bone tissue engineering, combinations of synthetic polymers, a 3D-porous composite of polylactide-co-glycolide (PLAGA) and 45S5 bioactive glass (BG) have also been tested [36].

Though it is clear that the mechanics of cell arrangement plays a crucial role in tissue architecture, as of now we understand very little about its dynamics and control [37]. Maintaining or anticipating a particular mechanical effect is challenging due to the fact that the mechanics of the overall system changes profoundly with the growth of cells. Even the expression of different proteins, especially those involved in holding the cells together in tissue format, can modulate the mechanical behavior of the system [38].

3.1.2.4 ECM mimicking through organ/tissue decellularization

Challenges in tissue specific mimicking of the ECM has given rise to a novel approach of decellularizing a spare functional organ/tissue, a top-down approach. The 'bottom-up' approach needs to mimic the shape of the targeted tissue/organ besides replicating the mechanical and biochemical characteristics of the native ECM to yield a physiologically functional tissue. Whereas the 'top-down' approach involves removing cells and leaving the native, original ECM framework that can be recellularized with appropriate cells. This approach could also help in different ways, for example in acquiring better understanding of functional ECM constituents and their relative orchestration in a normal healthy tissue. In this context it is pertinent to decellularize the tissue without damaging the ECM. Moderate experimental conditions and chemical treatments to dislodge the cells from their matrix are expected to spare the ECM structurally and compositionally intact. The mechanical properties and the nano- to macro-features of the remaining matrix could allow tracing the modular arrangement of major ECM components. This approach is particularly useful for cell-dominant vital tissues like heart, kidney and lungs where an ECM study is not easy otherwise. Unlike skin, bones, vascular tissues which are ECM dominant and also in relative abundance, these tissues are complex, limited by availability and dominated by indefatigable physiologically active cells.

Designing and shaping architecturally complex organs like the heart and kidney are difficult to replicate through scaffolds. Myocardium particularly is thick (~10 mm), dense, vascular and highly sensitive to ischemia [39]. Oxygen consumption by the cardiomyocytes is also much higher than on an average demand by other cell types. Engineering an artificial construct with unidirectional valves and architectural intricacies like the internal partition has led to such an alternative way of creating a true replica of the cardiac ECM through decellularization [40]. Re-endothelialization prior to recellularization of the left ventricle wall of the native cardiac scaffold through vessel and cavity could improve the contractility of the heart construct [9, 40]. ECM specific knowledge acquired through such explorations could help in designing at least small functional cardiac patches to replace and integrate dead or scarred tissue. Such functional micro-tissues, could also find application in ex vivo drug testing.

3.1.3 Need for Compatible Cell-Types

It has been learnt through the study of developmental biology and human physiology that each tissue comprises of multiple cell types that are compatible and mutually supportive. In fact it is only through their concerted activity that a tissue can perform a physiological function.

It is interesting to note that almost all the organs with distinct physiological activity consist of smaller tissue units that are independently functional. A heterologous population of cells that organize into uniformly patterned structural units acquiring distinct shape and physiological functionality is the key feature in them. Multiple such units, though variable in size, exist in an amenable relation with their surroundings and impact the physiological performance of the whole tissue. Each tissue is thus unique not only in its unit-morphology but also in its commitment to a specific physiological function. Islets of beta cells in the pancreas, glomerular units in the kidney, acinus in the liver, and seminiferous tubules in testes are some of the fine examples. These functional units may or may not be confined by membrane but remain tightly packed with supporting interstitial cells which remain closely aligned for feasible co-ordination. Non-linear but systematic functional co-ordination of this kind, among cells within each unit and with neighboring units demands connectivity through a responsive peripheral ECM. Concerted units in a defined tissue/organ format remain connected to other organs through vasculature that routes the paracrine and endocrine signals. Multiple functional units portray Nature's typical design advantage that allows efficient tissue function even in a state of accidental compromise, e.g., due to nutrients or energy shortage or other transient stress effects [12].

3.2 Models for 3D Cell Culture

3.2.1 Spheroids

Cells multiplying under shear forces often group together and form a lump or cell-aggregate called a spheroid. Non-adherent type of cells have a tendency to form spheroids while multiplying. In general spheroids are formed when cells are grown in spinner flasks, a technique developed in 1970 by Sutherland and his coworkers [41]. Sometimes they may also represent a three dimensional cell mass generated from a single cell in the epicenter. In such cases the cells multiply while pushing newer cells in the outward direction. Spheroid culture is preferred for testing anti-cancer and anti-tumor drug candidates not only due to their resemblance with tumor morphology but also due to the better accuracy it may offer with drug sensitivity assays. The morphological difference in the spheroid culture (3D) system in comparison to the flat bottom plate (2D) culture for 3 different cell types is depicted in Fig. 3.4. Classical methods of creating spheroids involve growing cells in suspension, in a round bottom non-adherent surface or as hanging drops [42]. Some of the special culture wares derived from the existing materials for growing spheroids are shown in Fig. 3.3.

Figure 3.3 Cell culture wares adapted to grow spheroids as hanging drops.

Using spheroid culture as a 3D model for detailed study is, however, hampered due to its delicate architecture. Stabilizing and maintaining the delicate spheroid architecture way through the study has been found very challenging. Closely associated multiple layers of cells also pose a limitation in imaging without disintegrating the spheroid. Detailed morphological examination of spheroids and its cells in a non-destructive manner is not easy.

Some of the existing challenges are addressed by preserving the spheroids through the silica bioreplication method developed by Lou et al. [43]. This method of spheroid preservation is based on the self-limiting biomolecular surface-directed silica assembly process that gives rise to a nearly accurate in-scale replica of external and internal cellular [44, 45] tissue and organism [46]. The process allows the biological sample to get imprinted in nanometer (<10 nm) thick silica layers. It was demonstrated that pluripotent 3D spheroids of human pluripotent stem cells (hPSC) could be recovered from the hydrogel by cellulase enzyme for detailed examination.

3.2.2 Hydrogels

Hydrogels are normally made up of hydrophilic polymers that bloat on absorbing water. Water incorporation results in the generation of hydrophilic and lipophilic pockets. Among different biomaterials, hydrogels are one of the potential candidates for TE and RM applications as they can mimic the physical, chemical, electrical, and biological properties of most biological *tissues* [47–49]. Hydrogels possess a three-dimensional (3D) highly hydrated polymeric network, and can hold up to 20- to 40-fold more water compared to their dry weight. Due to their unique physical properties, these

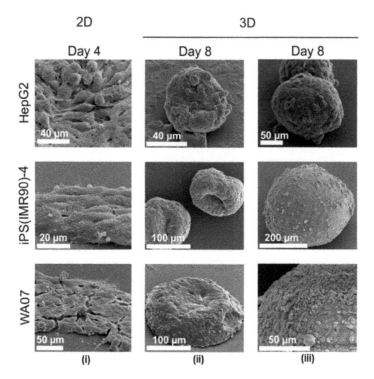

Figure 3.4 Morphology of human pluripotent stem cells and hepatocellular carcinoma cells with and without silica bioreplication. (a) SEM images of HepG2, iPS(IMR90)-4, and WA07 cell spheroids show deformation of 3D spheroids. (b) SEM images of HepG2, iPS(IMR90)-4, and WA07 cell spheroids after silica bioreplication. (Adapted with permission from Ref. [43].)

networks can be shaped or casted into various sizes and forms. The polymeric arrangement in hydrogels can be exploited for cell culture as such or could be stabilized through controlled cross-linking. The size of these pockets depends upon the nature of the constituting polymers. These inter and intramolecular pockets generate spaces that allow cells to expand within and sometimes over the surface as well. Hydrogels are also found useful for sequestering or adsorbing growth factors and other functional molecules persuading the growing cells in desired directions. Lou et al. developed a plant-derived nanofibrillar cellulose hydrogel that could maintain the pluripotency of hPSCs (human pluripotent stem cells) for 26 days. This offers a 3D culture system that supports pluripotency and could be used in research and development involving stem cells [50]. Since it is possible to integrate nanoparticles in the polymeric hydrogels, they provide scope for design engineering. Nanocomposites with superior physical, biochemical, structural and biological properties could therefore be designed to address biomedical and biotechnology needs [51].

3.2.3 Scaffolds and Matrices

Scaffolds represent a molecular network artificially created to mimic the spatial arrangement of native ECM in 3D. Either microporous or fibrous scaffolds may be used for 3D cell culture. A scaffold provides the physical framework in 3D space onto which cells are seeded. The fibrous material also needs to generate spaces to allow micron sized cells to penetrate and expand. Gibson and Ashby have classified porous solids into two groups, honeycombs and foams [52]. However, as per ASTM (American society for testing and materials) standards, porous materials can be classified into three categories: (a) materials with interconnecting pores, (b) closed or not interconnected pores and (c) a combination of both a and b. Porosity and its distribution in a scaffold is one of the parameters that impact the growing cells most. Pore size, pore walls, edges and their relative density have a major impact on scaffold mechanics. Table 3.1 below exhibits the influence of pore size and its distribution in the scaffold in the context of bone tissue engineering or bone-implant integration [53–56].

Any biocompatible polymer can be used for creating scaffolds. However, the biomaterial scaffold should be degradable by the

Table 3.1   Implications of scaffold pore size in bone tissue engineering

Pore size (μm)	Biological relevance
<1	Protein interaction and adsorption
1–20	Initial Cell attachment and directed cell growth
20–100	Cell proliferation, migration
100–1000	Cell growth and collateral bone growth
>1000	Maintaining the shape and overall functionality of implant

enzymes released from the cells to make way for growing cells. It is considered ideal if the kinetics of the degradation of scaffold biomaterial matches with the kinetics of the generation of neo-ECM by growing cells. This helps in maintaining the morphology of the cell mass in 3D while the supporting scaffold gets swapped by the neo-ECM. Though natural polymers by being more compatible are better, they are associated with limitations of availability, reproducibility and stability. Furthermore, being naturally originated they may carry immunogenic or virulent features and are also prone to microbial attack. Natural polymers are good biomaterials from the cell-interactive point of view, but their utility is also limited by their poor mechanical strength. For this reason the scaffold or matrices created from natural polymers most often end up forming soft hydrogels. As mechanical aspects of native ECM contribute towards tissue specificity through cell differentiation, soft hydrogels offer limited scope for tissue engineering purposes. In general there are four major considerations that need to be taken care of while designing a scaffold for tissue engineering purposes: (a) biocompatibility, (b) porosity (pore size, gradient distribution, volume fraction and interconnectivity); (c) mechanical (elastic modulus, compressive strength) attributes and (d) processing techniques based on the material (synthetic, natural or hybrid polymer).

In an attempt to mimic natural ECM, multiple, different synthetic and natural polymers are subjected to a variety of processing techniques (Table 3.2). Freeze drying, particle leaching, extrusion, melt molding, etc., are some of the conventional methods which essentially evolved to develop drug delivery systems but were adapted later for 3D cell culture. Yang et al. [57] have discussed some of these conventional methods while advanced methods are elaborated by Dutta et al. [58].

3.3 Materials for 3D Cell Culture

Biocompatible polymers, both of natural and synthetic origin are the preferred choice to create scaffolds or matrices for culturing cells ex vivo. However, co-polymers or polymers of hybrid molecules

Table 3.2 Some of the frequently used fabrication methods for creating scaffolds

Fabrication techniques	
Conventional	**Advanced**
(i) Solvent casting with particulate leaching	(i) Rapid prototyping (solid free form fabrication)
(ii) Extrusion	a. 3D plotting
(iii) Moulding	b. 3D printing
a. Compression moulding	c. Stereolithography
b. Melt moulding/blowing	d. Selective laser sintering
c. Injection moulding	(ii) Electro-spinning
(iv) Thermally induced gelation/ freeze-drying	(iii) In situ photopolymerization
(v) Gas foaming	(iv) High internal phase emulsion (HIPE)
(vi) Supercritical fluid processing	(v) Self-assembling peptides
(vii) Microsphere sintering	
(viii) Fiber bonding	
(ix) Sponge matrix imbedding	
(x) Fused deposition modeling	

generated by combining the natural with the synthetic molecules may also be used. Finding an appropriate scaffold material that can accommodate biological design variables inherently needed for each application is the key. Material selection is generally guided by the cell or tissue type to be grown. Each tissue type requires specific stiffness and flexibility in its surroundings. This aspect has already been discussed.

3.3.1 Natural

Proteins, glycoproteins, polysaccharides and their combinations which can represent an ECM-like environment may be used to fabricate the scaffolds for 3D cell culture. Among the natural polymers the most commonly used are proteins like collagen, gelatin and polysaccharides like agarose, cellulose, algenic acid, chitosan, etc. However, their source, functional and processing reproducibility are major concerns. Originating from live stock, they are prone to degradation by microbial attacks and therefore have a shorter shelf life. Further, they may continue to carry the infectious endotoxins and therefore unsafe as such for the purpose. Collagen, hyaluronan and algenic acid are among the most widely used naturally derived biomaterials. Hydroxyapatite, chondroitin sulfate, cellulose and silk are other natural materials invoking interest in the development of matrices and scaffolds. Collagen, a ubiquitously present constituent of the extracellular matrix is the most preferred and perhaps the most extensively explored. It is available in abundance and possesses unparallel structural strength due to its unique triple helical architecture. Hyaluronan, a distinctive disaccharide present in high concentrations in the aqueous humor of the eye, synovial fluid, skin and umbilical cord besides being widely distributed in low proportions in the extracellular matrix has also been experimented on to create scaffolds. Alginic acid or sodium alginate is another natural polysaccharide isolated from brown algae that has been tried. Being a linear copolymer with a high water retention capacity, it forms a good hydrogel.

Compared to synthetic polymers, natural biomaterials are found more compatible but limitations like endotoxins, viruses associated with the biological system and their unpredictable consequences may still be an issue. Nevertheless, technologists are now relying more on processed natural biomaterials, hoping that chemical processing would be able to neutralize/nullify the toxin related concerns and the rest could be managed through the sterilization of final products.

3.3.2 Synthetic

The utmost important condition for a synthetic polymer to find applications in the creation of scaffolds is their bio-compatibility. They should also be bio-degradable, without leaving any byproduct that may prove toxic to the cells/system. Synthetic materials and high density synthetic polymers due to their inherent strength have found applications in hard tissue engineering. Polyethylene glycol (PEG), polyvinyl alcohol (PVA), polycaprolactone (PCL), poly-lactic acid (PLA) and poly-glycolic acid (PGA) are some of the synthetic polymers that are found to be bio-compatible. Being immunologically neutral and biodegradable they are frequently used for fabricating medical devices. Their availability in a wide molecular weight range helps in controlling the elasticity of the device. Among the inorganic compounds commonly chosen to create prosthetics are calcium and strontium salts, hydroxyapatite, titanium and siliconized materials. These inorganic materials are neutral and in general do not incite any immune response. Porous ceramics and those toughened with alumina and zirconia are the new composite materials specifically being designed for bone tissue culture and engineering. However, devices made-up of these inorganic materials should have a smooth finish as even small grooves or roughness in the submicron scale may invite immune cell attacks.

Synthetic polymers and structurally well-defined co-polymers that can provide a range of cost-effective materials are widely preferred for creating scaffolds and matrices. This is simply because of the ease of processability [59]. Nonetheless, synthetic biomaterials often tend to degrade and release acidic by-products that can alter the pH of the surroundings. This could be a serious concern as it may lead to adverse tissue and inflammatory reactions [57].

3.3.3 Hybrid

Hybrid materials generated by a physical or chemical combination of two polymers of the same or different origin have also found a place in developing cell culture scaffolds and devices. The purpose of creating such blends is mainly to incorporate the desired physical, mechanical and biochemical properties. This means the alliance could be commenced using two or more synthetic molecules/polymers, two or more natural molecules/polymers or combining synthetic with natural molecules/polymers and so on. Experience and literature supports the fact that hybrid materials with such permutations and combinations have proven better in many ways. They impart flexibility and sometimes add a new dimension to the functional ability of the designed material. These hybrid materials could be achieved either by physical mixing and then processing them together or by creating a chemical bonding between the two or more materials of choice before processing.

3.3.3.1 Physical blends

Physical blends are generally used for creating bone tissue implants where hardness of the material is the major focus. Bone and cartilage tissue engineering particularly have considered hybrid biomaterials to tackle the demand of elasticity with the required inbuilt strength in the scaffold. Replacing bones with metal and steel-like rigid material for support has a long history. However, since metallic implants are unable to provide the requisite flexibility and elasticity that could match the biological tissues like cartilage and bone, the attempts were shifted to alloys and later to ceramics with advanced compositions and structures [60]. More recently, titanium and its nickel alloy called Nitinol is developed. It is light yet strong therefore proved one of the favoured inorganic blends for bone implants [61]. Blends of the hardest known materials like titanium, zirconium and other metallic elements in different proportions with synthetic polymers like polyethylene and polystyrene are also being tested. Generally, high density or ultra high density of synthetic polymers is required to yield a high strength uniform blend. It is now well understood that a material can perform better as an implant only if it is homogeneous. Heterogeneity at a gross macromolecular level may give in to the stress and even a tiny fracture could propagate and lead to implant failure in finite time. Ultra-high-density polyethylene is the universal norm today in matching the strength of metallic components in the blend. Carbon, barium and strontium blends are also being tested for bone implants. These elements are expected to impart bone-like piezoelectric properties to the graft material, making the implant functionally more accurate. Silver compounds and carbon nanotubes are particularly being pursued to introduce other advantages like antibacterial properties, etc.

Bone prosthesis/implants need to be integrated with the host system for long term benefits. This could be achieved through an interface using bioorganic scaffolds or biomaterials that can facilitate bone ingrowth. Exploring new avenues of strength combined with the desired elasticity through scaffolds generated from hybrid biomaterials is therefore a new trend in bone and cartilage tissue engineering. Chitosan-alginate [62] and PLGA-silk [63] polymer hybrids and inorganic-organic hybrids like silica-chitosan [64] have been recently evaluated for the creation of flexible scaffolds for bone tissue engineering. The chitosan-alginate hybrid scaffold is shown to be better with significantly improved mechanical and biological properties where bone forming osteoblasts readily attach, proliferate and deposit calcified matrix. Through in vivo study, Li et al. [62] have demonstrated the high degree of tissue compatibility and calcium deposition occurring as early as the fourth week after implantation. Another promising combination of polymers for cartilage tissue engineering is shown to be poly-ϵ-caprolactone with hyaluronan and chitosan [65]. Scaffolds made up of both these hybrid biomaterials promoted neo-cartilage formation as quantified by the dimethyl-methylene blue assay for sulfated glycosaminoglycans (sGAG). This was also confirmed by the expression of type II collagen through ELISA. The scaffold also exhibited initial cell retention.

Earlier, steel or metal implants were intentionally made smooth with a polished surface finish so that the implant might not attract host immune cells. However, lack of integration with the implant-site is now realized as the major cause of implant loosening and failure. With new biomaterials and their behavioural understanding in the biological system, integration with the surrounding tissue is recommended. Osteointegration is therefore considered the key to the success of a bone implant. An efficient integration with the host body could prevent loads of complex morbidities post surgery. This demands nanoscale roughness on the implant of fully compatible biomaterial, which is expected to allow a friendly alliance/association with the neighbouring healthy tissue through ECM. Nano- and microporous biomaterials, including organic-inorganic composites possessing bone-like strength are particularly useful in correcting bone defects. Nanoporous material with optimal strength is also being developed with the objective of bone cell culture and bone tissue engineering. Jose et al. [66] have recently reported a nanofibrous biomaterial scaffold based on PLGA and nanohydroxyapatite where fibers are aligned to restrict chain mobility and prevent shrinkage, making it further suitable in handling bone defects. Such scaffolds will be useful in creating a nanomicron scale coating on the bone-grafts/implants made up of the most widely used materials like titanium and its alloys.

Titanium-hydroxyapatite composites and their coatings may also find use in dental implants.

3.3.3.2 Chemical blends

Chemically bonded combinations have generally been explored to create scaffolds and matrices for the purpose of drug delivery, soft tissue engineering and regenerative medicine. The chemically combined molecules are expected to be superior in terms of biocompatibility or other functional properties that one seeks in the scaffold. That each individual component should contribute to and improve the scaffold performance is the basic guiding principle. Therefore, chemical reactions are performed to improve upon an existing property or to impart a new property to the resultant matrix. A co-polymer of poly-lactic and poly-glycolic acid (PLGA), for example is found superior in both biocompatibility and bio-degradability. As of now it is the most preferred synthetic co-polymer in developing sustained release formulations. With established biocompatibility, acceptable biodegradation rate, PLGA has already been approved by the U.S. FDA for clinical use. Furthermore PLGA grafts can easily be modulated for cell interaction which could improve their tissue integration. The possibility of altering its shape and surface properties through design principles makes it the material of choice in the creation of bone substitutes [67]. Similarly hybrid scaffolds composed of beta-chitin and collagen [68] and a flexible scaffold of silica-chitosan [64] with oriented pores have also been generated for 3D cell culture purposes.

3.4 Methods of Creating Scaffold/Matrices for 3D Culture

It is understood now that not only the material but how it is being processed to make it suitable for cell culture is important. The fabrication or processing technique has a great impact on the characteristics of the resultant scaffold. Innumerable innovative methods are being attempted world wide to create scaffolds while accommodating the desired features. The diversity of approaches explored for designing 3D scaffolds and matrices is in itself incredible. These techniques have evolved and have been guided mostly by the need and purpose of applications. Some of them have even borrowed from typically disconnected fields and have been adopted ingeniously for the purpose. For simplicity, these methods can be classified into two categories (i) conventional and (ii) advanced techniques. The conventional methods in general involve old, well established concepts and techniques which could be accomplished through existing tools and machinery. The advanced techniques on the other hand are those which incorporate either fresh innovations in existing tools or demand new ones based on ideas developed exclusively for the purpose. They may at times involve computer aided design (CAD) model adaptations of old methods. Some of the commonly used scaffold fabrication techniques are briefly described below.

3.4.1 Conventional Techniques

Traditional fabrication techniques like blending, sintering, etc., are adopted as such to create scaffolds mainly for hard tissue engineering. Bone and cartilage-like tissues which need cells to be grown on hard matrices involve materials like hydroxyapatite, metals and alloys that can withstand high temperature and pressure. They are generally simple one or two-step methods which are well established, even if for different purposes in different technical fields. Yang et al. have elaborated on the traditional methods of scaffold design [57, 69]. Some of these conventional techniques are listed below.

3.4.1.1 Solvent casting/melt molding

Porous scaffolds are successfully fabricated by melt-molding or solvent casting while combining them with particulate leaching [70]. Freeze-casting or ice-casting is also combined with other methods to introduce porosity or gradient porosity [71, 72].

3.4.1.2 Extrusion

Extrusion is the technique commonly used for imparting different shapes to molten polymer/plastic articles. Automation and precision introduced over time has encouraged its use in creating implants, scaffolds and matrices. Experience and CAD adaptations in the existing set-up has raised the possibilities of creating even complex designs through extrusion [69].

3.4.1.3 Molding with particle leaching

This technique involves a mold of desired shape and size. In order to create porous 3D scaffolds for cell culture purposes, the method is slightly modified. Here the biomaterial is mixed with particles soluble in some specific solvent (polar or non-polar). After compression molding into the desired shape, the biomaterial-particle composite is washed with a suitable solvent. The particulate material dissolves and leaches out, leaving behind gaps or pores of its size and morphology [73].

(i) *Compression molding*: Here shaping of the biomaterial is achieved through compression. The molding material is generally preheated and poured into a heated, pre-shaped cavity/mold. The mold is then closed with a top force so that the material reaches all areas of the mold and acquires its shape. The shaped material is then cured under suitable temperature and pressure. This process involves thermosetting polymers and resins in a partially cured stage either in the granules, pellet or putty formats.

(ii) *Melt molding*: Melt molding is the term that was used earlier for shaping a material through extrusion. In this technique the molten material is shaped while being extruded.

(iii) *Injection molding*: Injection molding is also a contemporary manufacturing process which is commonly used in producing products from thermoplastic or thermosetting plastic materials. It is also adopted to create scaffolds for 3D cell culture purposes [74].

3.4.1.4 Gel casting and freeze drying

Freeze casting [71] and a combination of gel-casting and freeze drying methods are also used for preparing porous hydroxyapatite scaffolds. This has helped in creating a scaffold microstructure with improved mechanical properties [75–77].

3.4.1.5 Thermally induced phase transition/gas foaming

The thermally induced phase transition [78] and gas foaming by compressing the biomaterial with the high pressure of CO_2 gas for creating open pores in the scaffold has also been tried [79, 80].

3.4.1.6 Fiber bonding

Fiber bonding technique involves heating two different nonwoven fibers in a manner to generate inter connected fiber networks in a predesigned mold to impart the scaffold a structural form or shape [81–83].

3.4.1.7 Microsphere sintering

It is a method where microsphere scaffold is created through heat sintering [72]. This technique is popular for fabricating ceramic or ceramic-incorporated polymer scaffold or matrices. It is useful for constructing substrates for bone and dental implants. It is also a favoured process for producing porous scaffold and matrices for bone tissue engineering mainly because of the mechanical strength acquired by the scaffold during processing [84].

3.4.1.8 Micro-contact printing

Micro-contact printing that allows patterning of covalently grafted polymers represent a soft lithographic technique [85]. It is possible to make 25 nm thick patterned films through micro-contact printing-based lithography [86].

3.4.2 Advanced Techniques

Advanced techniques to create scaffolds are generally novel with innovative origins. However, some of them are also derived from methods conventionally developed for drug delivery purposes [87]. Unlike conventional methods, they involve sophisticated tools developed for the specific purposes. Being new, these methods also present scope for optimization, customization and standardization. A few such techniques are described below.

3.4.2.1 Electrospinning

Electrospinning is a contemporary technology with wider application potential [88]. Thin micron sized fibers of synthetic [89] as well as natural polymers [69, 88] and their hybrids [90] can be produced through electrospinning. This method is found very useful in creating fibers of nano to micron-size thickness using synthetic polymers like poly lactic acid (PLA), poly glycolic acid (PGA), poly ε-caprolactone (PCL), etc. Ease of processability and feasibility to use organic solvents make it more compatible for creating fibers of synthetic polymers. It has been reported that keratinocytes and endothelial cells could proliferate and organize in native epidermal-dermal structures at the air-liquid interface even in the absence of serum when co-cultured with fibroblasts on an electrospinned polystyrene scaffold created by electrospinning. Such macroporous scaffolds of packed fibers of 10 micron size could yield self-organization of skin cells. This in fact paved the way for synthetic scaffolds that could be used for 3D or tissue-like cell culture in vitro [91]. Nevertheless, the inherent hydrophobicity of such synthetic materials may cause hindrance to cell attachment; blending these synthetic materials with natural polymers such as collagen or gelatin improves cytocompatibility [92]. PCL-gelatin (1:1) matrix scaffolds, for example are found better suited for bone marrow stem cell (BMSC) penetration as compared to pure PCL nanofibers. A silica-based sol-gel glass ($70SiO_2 \cdot 25CaP \cdot 5P_2O_5$) blended with a polymer binder was recently electrospun into a nanofibrous mesh and subsequently heat treated to produce fibers with varying sizes depending on sol concentration. The glass nanofibers supported the formation of bone mineral-like apatite phase on the surface in a simulated body fluid, possibly due to extremely large surface area of nanofiber and consequent ionic reaction with the surrounding medium. Moreover, a comparison of osteogenic differentiation of rat BMSCs on bioactive glass and a nanofibrous scaffold clearly illustrates the biological and morphological advantages associated with usage of bioactive glass nanofibers [93]. Preliminary cell culture studies on scaffolds created by electro-spinning are very encouraging. However, numerous iterations required to create a scaffold in the millimeter to centimeter size range makes this technique somewhat tedious and brings scale-up issues.

3.4.2.2 Rapid prototyping or solid free form fabrication

Rapid prototyping represents a group of methods that can generate a physical model directly from computer aided design data [94, 95]. Also known as solid free form fabrication technique, rapid prototyping allows accurate control over macro- and microproperties of the scaffold through CAD [96]. It is an additive process in which each part/component could be constructed layer by layer. Supported by computerized models of damaged tissues or bones from patients' MRI to facilitate the fabrication of the scaffold with precise size and design parameters, this technique is especially useful for bone repair. Three dimensional (3D) printing where fabrication is done through inkjet using polymer ink and 3D plotting, where the scaffold is created layer by layer are the two RP methods that allow the fabrication of very complex architectures. In the 3D powder printing method, a printer head is employed to spray a liquid on the thin layers of powder, which basically follows the object's profile generated through the CAD model. The parts are built up layer-by-layer, which are produced by a roller transferring powder from the feed chamber to the build chamber. In the next step, a binder is deposited locally by the printer head according to the actual cross-section obtained from the CAD model [95]. This sequence of steps is iterated until the desired part is fully fabricated. The fabricated structure is surrounded by unbound powders which are finally removed or blown away from the printed model. This technique of 3D powder printing is capable of fabricating parts using a variety of materials, including ceramics, metals and polymers with diverse geometries. Rapid prototyping which includes 3D printing and plotting offers flexibility to accommodate cells since it involves mild operating conditions. The genesis of 3D plotting lies in the field of biomedical science unlike other RP methods which were adapted from mechanical engineering disciplines. In 3D plotting, a polymeric paste with the desired viscosity can be processed at a desired temperature.

Rapid prototyping also makes it possible to create pore size gradient in a single step. The pore-size gradient scaffolds improve seeding efficiency from ~35% to ~70% in homogeneous scaffolds under static culture conditions [97]. However, both these techniques are more useful only if the involved organic chemicals are compatible with nature- derived bio-polymers. Similarly, laser sintering and fused deposition techniques, where scaffolds are created by sintering or fusing porous particles in a powdered state, can create a scaffold of any complex shape but may not be as useful with biopolymers due to high processing temperatures. Leong et al. have meticulously tabulated the advantages and disadvantages of some of these rapid prototyping techniques [98].

(i) *Stereolithography/3D laser lithography*: Stereolithography is a process of creating 3D objects by using computer controlled laser beams. Utilizing a 3D image, a part is built slice by slice from bottom to top in a vessel of liquid polymer that hardens when struck by a laser beam. The method is also useful in creating scaffolds with gradient porosity [99].

(ii) *CAD-based 3D plotting*: Mainly useful for bone-implants computer aided design of the scaffold [95] has enabled the creation of customized grafts based on the defect site. Complex structures in the micro- and macroscale could be created for a scanned and digitized image of the defect site [100]. A number of CAD-based interface programs have been developed to achieve the bidirectional conversion between commercial CAD systems and the neutron transport simulation codes [101]. The success of this approach depends on the accuracy and retrieval of visual similarity–based models. Simultaneous progress in the software for acquisition and reproduction of 3D images is helpful in using digital information for devising customized implants and scaffolds.

(iii) *CAD-based 3D printing*: 3D printing is a rapid prototyping technique where biomaterial is used as ink to print the scaffold in three axial directions in a controlled manner in order to yield the desired shape and size [102]. The technique is found suitable especially for creating custom implants based on medical data sets [103]. 3D printing allows fabricating scaffolds for bone defects using hydroxyapatite (HA) with complex internal structures and high resolution [104].

(iv) *Fused fiber molding*: In this method the scaffold is created by taking the polymer or a mixture in a heated liquefier and then pouring or extruding in its fused form through a fine needle. Material flow is regulated by applying pressure to the syringe. The extruded fibers or drops are directly deposited on a plate or rolling drum at normal or low temperature that helps the material to harden and take the shape [105, 106].

(v) *Fused deposition molding*: Also known as fused deposition molding, another rapid prototyping technique where physical model is prepared by depositing layers of thermoplastic material one at a time [107]. The molten material is extruded through heated nozzle and deposited as thin solid layers on a platform. FDM allows reproducible production of scaffolds with interconnected pores and is very useful for bone tissue engineering applications where hard and heat stable materials could be used.

(vi) *Selective laser sintering (SLS)*: Selective laser sintering or laser sintering is an additive manufacturing process that uses lasers to fuse or sinter powdered material (typically metal and thermoplastics). The technique allows creating geometrically intricate designs and since the 3D printed object remains embedded in the constituting powder material it eliminates the need of post treatment for keeping the structure intact [108]. It is an evolved version of microsphere sintering a conventional method where sintering is done through high temperature heating.

The technique is adopted as such to fabricate clinical implants using ultra high molecular weight polyethylene (UHMWPE) [109, 110]. SLS-fabricated calcium phosphate (CaP) implants are reported to be useful [95]. Laser sintering is also used for creating porous scaffolds by sintering microspheres [111]. In order to accommodate temperature sensitive materials, the sintering of microspheres through sub-critical CO_2 is also attempted [112].

3.4.2.3 Emulsion templating

Emulsion templating for scaffold fabrication is a method adapted from the non-biological field. It was developed for polystyrene synthesis and has found its way into many different fields of engineering including fabrication of scaffolds for 3D cell culture. Emulsion templated porous polymers are the porous polymers synthesized within high internal phase emulsions, which was first described in 1962 in the context of polystyrene synthesis [113]. The application potential of PolyHIPES as reaction supports, filtration membranes, responsive and sensing smart materials, scaffolds for control release and tissue engineering, and templates for porous ceramics and porous carbons is exciting. The last decade has witnessed a rise in the exploration of polystyrene [114] and acrylic acid [115] based HIPES for their potential use in tissue engineering. Hayward et al. have recently reported galactose functionalized PolyHIPE for the routine 3D culture of hepatocyte [115].

3.4.2.4 Micromolding

Micromolding is the technique which combines precision with injection molding to create articles with microscopic accuracy. Successful development of ultra-high density polymers with much improved strength has led to its use in almost all spheres of life. Plastics being lighter if incorporated with strength comparable to metal could easily become a preferred replacement. The chemically inert nature of plastics also makes them superior to metals for multiple applications where insulation is required. Possibility of creating microstructures with precision has made it convenient to create micro-featured components for all types of articles including household appliances, toys, medical devices, vehicles, aircrafts, ships and even spacecrafts and satellites.

Micromolding is also used for creating PDMS scaffolds and cell culture devices [116]. Multilayer micromolding is employed in fabricating 3D polycaprolactone (PCL) scaffolds [117]. Recently a well-defined macroscopic scaffold was created using two-photon polymerization and explored for neural tissue engineering [118].

3.4.2.5 Photoplating/photolithography

The photopolymerization technique to create a 3D scaffold was developed as a consequence of newly identified photopolymerizable polymers [119]. This technique involves optical energy to irradiate and fuse the thin layer at the surface of a liquid photopolymer resin. Polymerization happens due to a functional moiety which is sensitive to light at a specific wavelength. Such chemical change on exposure to particular electromagnetic radiation renders the material to transform into a solid phase [120]. Relevant functionality can be chemically introduced in a desired polymer for the creation of designed biomaterial.

3.4.2.6 Designed self-assembly

Molecular combinations that have a tendency to aggregate owing to their complementary charge offer a possibility to convert into a matrix. Polymeric chains created from such molecules allow control over the designed matrix formation at the molecular level. Self-assembling peptides [121] are one such macromolecules that have been explored to create designed biomaterial scaffolds [122]. Chemical complementarities and structural compatibility through non-covalent interactions are the key elements in such self-assembling molecules. A good understanding of these elements may help to design and manipulate the morphological features of scaffolds at the molecular level. Since structural-functional features of peptides could be modulated by varying the amino acid sequence and composition, the fabrication of architecturally diverse kinds of scaffolds is possible [123, 124]. If designed appropriately, such peptides and polymers could offer a unique possibility of in situ assembly through the injection mode. Thermo and photo-sensitive self-assembling peptides are some of the very innovative recent explorations in generating matrices [125]. Such biomaterials can also be rendered tissue-specific by incorporating relevant signaling molecules or growth factors without compromising their ability to self-assemble. Although this approach offers design flexibility at the molecular level, the lack of mechanical strength limits their application potential. Peptides normally form hydrogels that are too soft to provide support to the cells.

It is clear from the above discussion that scaffolds for 3D cell culture could be prepared by a number of methods (Table 3.2). Mostly inspired by conventional techniques, integrated new innovative ideas have helped to develop different options for scaffold

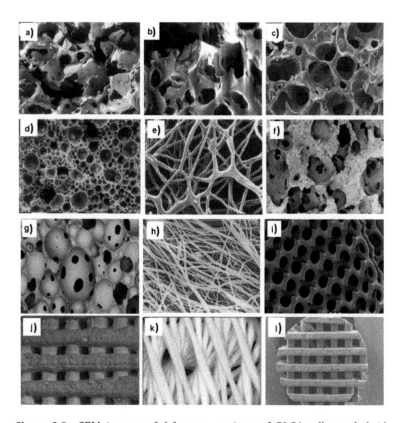

Figure 3.5 SEM images of (a) cross-sections of PLGA-collagen hybrid sponge-like scaffold fabricated by particulate-leaching method (*Chen et al., 2002); (b) silk scaffold prepared by salt leaching and gas foaming methods after immersing it in butanol (Nazarov et al., 2004); (c) PLA polymer after processing in super critical CO_2 at 240 bar, 35°C with 60 min venting (Quirk et al., 2004); (d) cross-section of 15% PLLA foams prepared by thermally induced phase separation (Nam and Park, 1999); (e) PGA mesh prepared by fiber bonding by heating (Mikes et al., 1993); (f) cross-section of scaffold prepared by sponge matrix embedding and (g) cross-section of scaffold prepared by foaming (Mastrogiacomo et al., 2006); (h) SEM image of 7.3% PLLA/1,2-dichloroethane scaffold prepared by electrospinning (*Zeng et al., 2003); (i) SEM images of PDLLA structures built by stereolithography (*Jansen et al., 2009); (j) calcium silicate ($CaSiO_3$) scaffold prepared by 3D printing (Wu et al., 2012); (k) porous scaffold prepared by fused deposition modeling (Kalita et al., 2003); (l) PCL scaffolds produced by the bioplotter with basic architecture (Yilgor et al., 2008).

fabrication. An interdisciplinary overlap and understanding has also led to the possibility of many different permutation combinations of known processing methods. Keeping the range of objectives both in vitro and in vivo in consideration, the quest for new scaffolds is still on. Macro- and microscale porosity in the scaffold is desired for tissue-like cell culture. Therefore porous and fibrous are the two basic constitutive morphologies inherently sought after by the bulk processing methods. Depending upon the tissue, gradient porosity is also found effective. Figure 3.5 exemplifies the morphological features of scaffolds created by different methods. Methods like SLS are operated upon layer by layer and are therefore tedious and time consuming. However, they allow control over morphology, the extent of porosity, biochemical topology and the mechanical properties of a scaffold as per the targeted need.

It is clear that the choice and compatibility of technique with the material is an important consideration. This depends heavily on the objectivity of the scaffold/matrix construction and also demands a unique system. It is interesting to note that the same material with

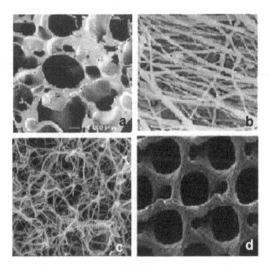

Figure 3.6 Scaffold architecture (SEM) acquired by PLLA depending on techniques of fabrication (a) phase separation-particulate leaching (*Chen et al., 2002); (b) electrospinning (*Zeng et al., 2003); (c) thermally induced gelation, solvent exchange and freeze drying (Yang et al., 2004) and (d) stereolithography (Jansen et al., 2009).

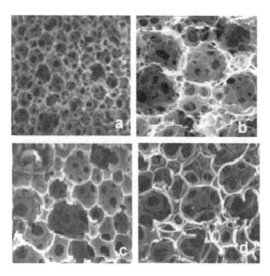

Figure 3.7 SEM pictures of look-alike scaffolds of (a) silica (SiO_2) (Zhang et al., 2004); (b) alginate (Barbetta et al., 2009); (c) PVA (poly-vinyl alcohol) (Colosi et al., 2013) and (d) gelatin (Barbetta et al., 2010) created by different manufacturing techniques.

different techniques can result in morphologically different scaffold or matrices (Fig. 3.6). Not only that, these refined techniques also allow chemically and constitutively different materials to produce architecturally similar and sometimes even identical scaffolds (Fig. 3.7).

Advanced fabrication techniques offer a great deal of control over scaffold properties. Self- assembling peptides for example can simulate the natural ECM architecture which originates from the hierarchical arrangement of nanoscale biopolymers [126]. Nevertheless as of now there is no technique efficient enough to be compatible with all kinds of material. In spite of many innovative approaches, the desired properties for a successful 3D scaffold may be of conflicting nature and difficult to be incorporated through a single fabrication technique. More often any adjustment or alteration in one property or feature would cost or compromise another property of the scaffold. For example, increasing porosity may influence density and strength of the scaffold and increasing stiffness through compression or crosslinking respectively may lead to altered density and hydrophilicity. None of the material and/or its processing method allows complete freedom of operation required for customizing a scaffold as the morphological, mechanical and biochemical needs of different tissue types are diverse. The technique should also be suitable for marketable operation, including scalability and automation to achieve high quality consistency at an affordable cost. Therefore no technique or scaffold is said to be sufficiently versatile for universal use. They would always need some modifications and adjustments specific to the target application [58].

3.5 Applications of 3D Scaffolds/Matrices

The cell-interactive 3D scaffold could potentially find applications in different spheres of health and medical related research and development. At present ECM mimicking has been successfully explored for limited purposes like enhancing the cell adhesive capacity of scaffolds through RGD and/or enzymatically degradable domains [127], culturing keratinocytes for skin equivalents [128], vascularisation [129] and hepatic tissue engineering [130]. Observations like the increase in collagen production by vascular smooth muscle cells (SMCs) when seeded within the scaffolds of PEG-based hydrogels [131] and adhesion and growth of vascular SMCs but not fibroblasts or platelets [132] by hydrogels modified with elastin-derived adhesive peptide are encouraging in terms of achieving growth through ECM management. However, an appropriately constituted ECM mimicking technology for 3D culture, which may allow customization for the growth of specified cells, could revolutionize our understanding of cell-ECM dynamics. With reproducible scaffold features and surface topography, an ECM mimicking matrix that qualifies for the necessities of strength, porosity, stiffness, and degradation dynamics will find applications in many diverse areas (Fig. 3.1), some of which are as follows.

3.5.1 In Research and Development

The major application of 3D matrix is in drug discovery as 3D cell culture tool. A 3D scaffold modified with ECM molecules/inducers can be used effectively as an R&D tool. Classical cell biology approaches are generally not amenable to the quantitative analysis of integrin receptor modulation and subsequent signaling and downstream cell responses [133]. Systematic investigation of signaling from ECM components therefore demands a model system where density and spatial organization of such ligands can be controlled. It is observed that cells placed in contact with a 3D scaffold attach rapidly, exhibit high plating and cloning efficiencies, proliferate quickly, reach a high saturation density, exhibit lower requirements for serum and added growth factors, respond better to hormones, express differentiated functions, have a longer life span, undergo morphological changes like flattening, and have better plating consistency. A 3D scaffold induces changes in cell shape and cell cluster arrangement not observed in cells grown in 2D on flat plasticwares. With such obvious advantages of growing cells in 3D, the outcomes of cell-biology studies using 3D-ECM substitutes are expected to be far more accurate.

3.5.2 Diagnostics

Cell culture–based assays are in common practice, especially to establish the extent of pathogenicity in diagnostics. They are also used as standards for efficacy/toxicity evaluation of various drugs. Intestinal absorption (using Caco-2 cell line), drug metabolism and drug–drug interaction (using primary hepatocyte culture) are some of the frequently used assay systems. Cells used for such assays could either be normal (primary) or transformed (cancerous) types. Unfortunately, the passaged cells are not the true representative of functional cells of the body, as the methods used for in vitro manipulation rarely constitute the natural in vivo environment that these cells are normally subjected to. Cancerous tissues particularly are heterogeneous and contain multiple subpopulations of cancer cells that differ in functional properties such as growth rate, metastasis and drug sensitivity [134]. Moreover, the isolated subpopulations (through 2D monolayer culture in Petri plates) do not reflect the complex nature of the tumor community and therefore their responses to test chemicals often lead to wrong conclusions. The natural response of isolated cells may not remain the same (as in the native environment) on exposure to conventional laboratory tools and systems. Poor diagnosis and therapeutic approach to cancer is thus appropriately attributed to the inadequate in vitro cancer models that use alienated culture condition to represent the disease [135]. Porous ECM mimicking scaffold-based 3D culture assays are therefore expected to bring a huge difference in diagnostics. Culture wares coated with appropriate 3D scaffold biomaterial are expected to improve the precision of diagnosis significantly.

It has been established now that the progression of malignancy from solid cancers is dependent on cell-cell interactions through entangled communication between cancer cells, cancer stem cells and stromal cells and also on the cell-ECM interaction [136]. Attempts have been made recently to recapitulate the complexity of the tumor microenvironment by incorporating immune cells. Monocytes, cancer associated macrophages, dendritic cells, lymphocytes and neutrophils are incorporated for the creation of immunocompetent 3D cancer models [137]. Biomimetic systems for investigating the pathophysiology of skeletal muscles are also being attempted [138].

3.5.3 Cell-Based Sensors

Having multiple receptors on their surface, cells depend on complex non-linear information processing that allows them to respond and identify some of the entities with supreme accuracy. This culminated into the idea of developing cellular sensors. Specific cell types are sensitive to specific signals and could find use in creating such sensors. Sensors for chemical and biological analytes, including pathogens, could prove useful in clinical diagnostics, food monitoring and detection of bio-warfare agents. Approaches like fluorescence energy transfer between pairs of chromophores or cells transfected with green fluorescent protein engineered to give a spectroscopic signal in response to a specific signal-transduction pathway may be used for sensing a particular agent in the vicinity of cells. Cells joined through a material device capable of translating natural signal transduction mechanisms to electrical outputs for detection could also be an approach. Cell interactive ECM scaffolds where cell surroundings can be manipulated to magnify their signaling response to the detection level can be of great help in achieving this goal. The piezoelectric behavior of bones which is particularly due to hydroxyapatite (HA) like crystalline components of its ECM may also be exploited to create sensors.

3.5.4 High Throughput Screening

Multi-well plates coated or inserted with a 3D-ECM scaffold could provide an easy means to perform high-throughput screening. After standardizing conditions for healthy proliferation and differentiation of a given cell line in a 3D environment, the intervention by drug molecules can also be evaluated.

3.5.5 Biotech Industry

Mammalian cell culture is an extensively utilized technique in the biotechnology and biopharmaceutical industry. Other than vaccines, a variety of enzymes, nucleic acids, hormones, interferons and antibodies are produced using suspension cell culture. Anchorage dependent cell lines like Vero (African green monkey kidney) and MDCK (Madin-Darby canine kidney) are the most frequently used cell lines for mass production [139]. Growth of these cells depends on the surface available to them for attachment and this led to the development of new culture wares and vessels with expanded surface areas. Porous 3D scaffolds with an appropriate pore size and mechanical strength can therefore find application in the upstream processing of vaccines and other biotech products where yield is directly correlated with the cell mass. Since the yield per batch is directly proportional to the surface area available for anchoring the multiplying cells, their use is expected to enhance the yield multifold. ECM mimicking of the scaffold to provide a biochemical microenvironment conducive to those specific cells could further improve the health of the produced cell mass. However, such an enhanced cell mass in a small volume demands extra oxygen and nutritional exchange that needs to be taken care of through special systems. The customization of a scaffold by integration with suitable bio-chemical entities for affinity extraction/entrapment could make it suitable for downstream processing as well.

3.5.6 Drug Delivery

An appropriately designed biodegradable 3D scaffold may also find use as a drug delivery vehicle [19]. Scaffold constituents may be chosen in such a manner that they could merge with the ECM at targeted sites while slowly releasing the contents/drug carried in the scaffold.

3.5.7 Biochemical Replacement

An ECM mimicking cell interactive biomaterial may also be used in hormone/enzyme replacement therapy. If attached with targeting moiety such scaffolds may also allow the supplementation or replacement of a specific cell mass, e.g., insulin-producing cells can be targeted or implanted in the pancreas. Alternatively, subcutaneous implantation of insulin-producing cells grown in the scaffold with an ECM milieu of islets of Langerhans may be used for diabetics where they can function as a natural reservoir for slow release or release on sensitization. Encapsulated hepatocytes for in vivo metabolic support may also be used in the same manner.

3.5.8 In Tissue Engineering (TE)

Regeneration and production of functionally viable tissue ex vivo is the ultimate goal in TE. The human body comprises of about 100 trillion cells, with about 260 different phenotypes that divide, differentiate and self-assemble over time and space into an integrated system of tissues and organs [140]. Therefore, precise replication of a functionally viable tissue is not a simple task. Successful TE involves a complex but regulated set of events that include the ability to accurately predict the cell response, acquisition of appropriate cells in adequate proportion and their nurturing in such a manner that they proliferate and differentiate into the desired functional phenotype. Engineering ex vivo humanoid tissue models and artificial tissues for implant are not réalisable without 3D cell culture. They demand not only architectural but tissue specific biochemical and mechanical mimicking of the ECM. Thus, tissue specificity of the scaffold in terms of native ECM constituents in a specified amount and state is another important requisite for tissue engineering. An ECM mimicking biomaterial for 3D cell culture needs to define and maintain 3D space to accommodate growing tissue besides assisting it to re-establish the in vivo like physiological milieu. A successfully engineered, functionally viable cell mass can be implanted to restore the functions of a lost or deficient tissue only if it can be integrated with the host body through a matching ECM biomaterial and vascular supply [141, 142]. The ECM mimicking scaffold can find use in TE in the following different ways.

3.5.8.1 Ex vivo organ model

To model organogenesis, i.e., the normal process by which cells self-assemble into functional tissues and then organs, is a big challenge. Knowledge gaps in the programmed differentiation pathways undertaken by different cell types in a specific tissue and their undefined interactive correlation further adds to the challenge. However, in simplified terms a tissue can be generated by co-culturing two or more types of cells at a specific differentiation state at close vicinity in a 3D environment, i.e., multi-cell-type culture. Certain tissues like skin and cartilage are easy to be recreated in vitro, as they involve only 2 or 3 types of cells, the differentiation of which is interdependent. However, it is expected that the 3D co-culture arrangement can replicate the dynamism existing between the cells and surrounding ECM more effectively. A study by Caspi et al. [143] supports this hypothesis. They observed that the co-culture of cardiomyocytes with epithelial cells led to appropriate organization only when fibroblast is added to make it a tri-culture. Thus, the cooperation and coordination through cross-talk among the parallelly growing cell-types apparently contributes enormously in recreating the tissue type. Such co-cultured, ex vivo organ models will also make it amenable to evaluate a potential drug candidate on specific cells in a tissue-like milieu. Innumerable experimental manipulations are possible with 3D organ type cultures as they can be used to elucidate the novel or re-examine the molecular signaling pathways characterized previously by conventional culture methodologies [144]. The ex vivo organ models can also provide molecular understanding of inter-cellular regulatory correlation in a tissue-like situation. Artificial skin-constructs, engineered ligament and tendon, engineered cornea/lens and substrates for nerve regeneration and periodontal use are some of the representative co-cultured products in different stages of development.

3.5.8.2 Tissue explants

Engineered tissue explants from autologous (host's own), allogenic (donor's), transgenic (same species), and xenogenic (different species) may be used for implantation. Cells can also be resourced from immortalized cell lines or stem cells (fetal or adult derived) depending upon immunological and other safety considerations of the application. However, the precise definition and control of cell potency is the major issue in such attempts.

Products for cardiovascular repair/regeneration, blood sub-stitutes and encapsulated cells for restoration of tissue/organ functions are some of the common tissue explants. Cartilage regenerating cells or bone repairing systems and bone graft substitutes are different types of tissue explants. Ex vivo systems like liver assisting devices for metabolic support are also considered as tissue explants [145] for which the ECM mimicking scaffold will be of good use. Advancements in the understanding of tissue specific requirements in a scaffold are simultaneously being translated and tested in creating artificial tissues ex vivo. For example, collagen and proteoglycan/glycosaminoglycan-based material could be tailored to achieve the microenvironment for primary skin cell culture [146]. The bilayer composed of a silastic epidermis and a porous collagen-GAG could be used successfully as a temporary cover in the cases of extensive burn injury (50–90% body surface area) where the patients have limited skin available for autografts [147].

3.5.8.3 In vivo tissue regeneration

Typically, an engineered tissue (e.g., skin) is created by harvesting cells from the patient, expanding them in culture followed by seeding them onto a scaffold. The cells on scaffolds are further cultured under optimized conditions to imitate a normal tissue-like growth morphology and differentiation before implantation. This culture period is often carried out in a bioreactor. A new technology where the biomaterial encapsulation could be constituted in situ along with the injected seed cells [148]. If successfully targeted it, would be a big leap in the direction of regenerating/supplementing the damaged/diseased tissues. By providing the quickest, easiest, highly compatible and least invasive solution such technology could prove obliging in many ways. It would require few cells and need no extra time for culture expansion. With limited in vitro exposure and the minimum need of laboratory chemicals, the chances of failure due to contamination would also diminish. It is worth mentioning here that a technique for seeding cells while crafting scaffolds of acrylated PEG by photo-polymerization for ex vivo applications has already been explored [149] and [150].

3.5.9 Human-Organoid Models for Physiology

The clinical evaluation of drug candidates largely depends on serial trials in rodents, monkeys and finally humans. Unfortunately, the genetic makeup and the finer physiological differences in the successive models forbid drawing an accurate inference regarding the efficacy and side effects caused by the drug. Developing ex vivo human tissue models could help in studying the effect of drug candidates directly on the human tissue analog, reducing our dependency on experimental animals. A successfully developed replica of human small intestinal villi using collagen hydrogel on a PMMA (poly-methyl methacrylate) mold [151] which provides a better tool for absorption studies is a good example. These 3D villous models with Caco-2 cell lines are found to exhibit cellular differentiation and absorption quite similar to mammalian intestines [152].

Co-culture systems could also be helpful in generating a tissue-like cell mass with a physiological resemblance. Two or more compatible and complementary cell lines need to be chosen based on experience with natural organ/tissue. The feasibility of generating a cell mass of different tissue types in the same culture plate would also authorize the study of relative impact of test molecule/material on more than one organ. By co-culturing two or more types of cells in the vicinity their influence on each other's behavior can be studied effectively (Fig. 3.8). Evaluation of cell behavior in a co-culture system where 3D-ECM-scaffold

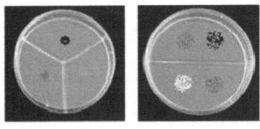

Co-culture with selective diffusion

Co-culture with free diffusion

Figure 3.8 Culture plates with 3D scaffold dots for co-culture.

beads/pellets seeded with different cell-types are kept in a direct or indirect vicinity while being immersed in the same culture medium could provide valuable information (Fig. 3.8). Depending on the objectives of the study, the compartments culturing different types of cells may or may not be partitioned. The possibility of pharmacokinetic and pharmacodynamic studies in such co-cultured ex vivo human organ models could provide an additional tool for the systematic appraisal of long-term clinical effects caused by chronic use of a drug.

References

1. Schmeichel KL and Bissell MJ (2003). Modeling tissue-specific signaling and organ function in three dimensions. *J Cell Sci*, **116**(12), 2377–2388.
2. Mann BK, Gobin AS, Tsai AT, Schmedlen RH and West JL (2001). Smooth muscle cell growth in photopolymerized hydrogels with cell adhesive and proteolytically degradable domains: synthetic ECM analogs for tissue engineering. *Biomaterials*, **22**(22), 3045–3051.

3. Gobin AS and West JL (2003). Val–ala–pro–gly, an elastin-derived non-integrin ligand: smooth muscle cell adhesion and specificity. *J Biomed Mater Res Part A*, **67A**(1), 255–259.

4. Dutta RC and Dutta AK (2012). ECM analog technology: a simple tool for exploring cell-ECM dynamics. *Front Biosci (Elite Ed)*, **4**, 1043–1048.

5. Koseki N and Yoshizato K (1994). Collagen-induced changes in the pattern of protein synthesis of fibroblasts. *Cell Adhes Commun*, **1**, 355–366.

6. Roskelley CD, Desprez PY and Bissell MJ (1994). Extracellular matrix-dependent tissue-specific gene expression in mammary epithelial cells requires both physical and biochemical signal transduction. *Proc Natl Acad Sci USA*, **91**, 12378–12382.

7. Barcellos-Hoff MH, Aggeler J, Ram TG and Bissell MJ (1989). Functional differentiation and alveolar morphogenesis of primary mammary cultures on reconstituted basement membrane. *Development*, **105**, 223–235.

8. Engler AJ, Sen S, Sweeney HL and Discher DE (2006). Matrix elasticity directs stem cell lineage specification. *Cell*, **126**, 677–689.

9. Robertson MJ, Dries-Devlin JL, Kren SM, Burchfield JS and Taylor DA (2014). Optimizing recellularization of whole decellularized, heart extracellular matrix. *PLoS One*, **9**(2), e90406.

10. Li WJ, Danielson KG, Alexander PG and Tuan RS (2003). Biological response of chondrocytes cultured in three dimensional nanofibrous poly(epsilon-caprolactone) scaffolds. *J Biomed Mater Res A*, **67**(4), 1105–1114.

11. Ng KW, Khor HL and Hutmacher DW (2004). In vitro characterization of natural and synthetic dermal matrices cultured with human dermal fibroblasts. *Biomaterials*, **25**(14), 2807–2818.

12. Dutta RC and Dutta AK (2016). Human-organoid models: accomplishments to salvage test animals. *J Biomed Eng Med Devices*, **1**, 110.

13. Choi CK, Breckenridge MT and Chen CS (2010). Engineered materials and the cellular microenvironment: a strengthening interface between cell biology and bioengineering. *Trends Cell Biol*, **20**, 705–714.

14. Orlando K and Guo W (2009). Membrane organization and dynamics in cell polarity. *Cold Spring Harb Perspect Biol*, **1**, a001321.

15. Janson IA and Putnam AJ (2015). Extracellular matrix elasticity and topography: material-based cues that affect cell function via conserved mechanisms. *J Biomed Mater Res Part A*, **103A**, 1246–1258.

16. Berthiaume F, Moghe PV, Toner M and Yarmush ML (1996). Effect of extracellular matrix topology on cell structure, function, and physiological responsiveness: hepatocytes cultured in a sandwich configuration. *FASEB J*, **10**, 1471–1484.

17. Drury JL and Mooney DJ (2003). Hydrogels for tissue engineering: scaffold design variables and applications. *Biomaterials*, **24**, 4337–4351.

18. Dutta RC and Dutta AK (2012). ECM analog technology: a simple tool for exploring Cell-ECM dynamics. *Front Biosci (Elite Ed)*, **4**, 1043–1048.

19. Dutta RC (2011). In search of optimal scaffold for regenerative medicine and therapeutic delivery. *Ther Deliv*, **2**(2), 231–234.

20. Wang N, et al. (2001). Mechanical behavior in living cells consistent with the tensegrity model. *Proc Natl Acad Sci USA*, **98**, 7765–7770.

21. Shenoy A and Shenoy N (2010). Dental ceramics: anupdate. *J Conserv Dent*, **13**(4), 195–203.

22. Pilliar RM, Filiaggi MJ, Wells JD, Grynpas MD and Kandel RA (2001). Porous calcium polyphosphate scaffolds for bone substitute applications – in vitro characterization. *Biomaterials*, **22**, 963–972.

23. Rizzi SC, Heath DJ, Coombes AG, Bock N, Textor M and Downes S (2001). Biodegradable polymer/hydroxyapatite composites: surface analysis and initial attachment of human osteoblasts. *J Biomed Mater Res*, **55**, 475–486.

24. Giannoudis PV, Dinopoulos H and Tsiridis E (2005). Bone substitutes: an update. *Injury*, **36**(Suppl 3), S20–27.

25. Liu H and Webster TJ (2007). Nanomedicine for implants: a review of studies and necessary experimental tools engineering most other tissues require scaffold with mechanical strength ranging from semifluid to gel. *Biomaterials*, **28**, 354–369.

26. Gleeson JP, Plunkett NA and O'Brien FJ (2010). Addition of hydroxyapatite improves stiffness, interconnectivity and osteogenic potential of a highly porous collagen-based scaffold for bone tissue regeneration. *Eur Cell Mater*, **20**, 218–230.

27. Shahini A, Yazdimamaghani M, Walker KJ, Eastman MA, Hatami-Marbini H, Smith BJ, Ricci JL, Madihally SV, Vashaee D and Tayebi L (2014). 3D conductive nanocomposite scaffold for bone tissue engineering. *Int J Nanomed*, **9**, 167–181.

28. Matsuno T, Hashimoto Y, Adachi S, Omata K, Yoshitaka Y, Ozeki Y, Umezu Y, Tabata Y, Nakamura M and Satoh T (2008). Preparation of

injectable 3D-formed β-tricalcium phosphate bead/alginate composite for bone tissue engineering. *Dent Mater J*, **27**(6), 827–834.

29. Gibson LJ and Ashby MF (eds) (1997). *Cellular Solids: Structure and Properties*. Cambridge University Press, Cambridge.

30. Dado D and Levenberg S (2009). Cell–scaffold mechanical interplay within engineered tissue. *Semin Cell Dev Biol*, **20**, 656–664.

31. Bell E, Ivarsson B and Merrill C (1979). Production of a tissue-like structure by contraction of collagen lattices by human fibroblasts of different proliferative potential in vitro. *Proc Natl Acad Sci USA*, **76**, 1274–1278.

32. Harris AR, Peter L, Bellis J, Baum B, Kabla AJ and Charras GT (2012). Characterizing the mechanics of cultured cell monolayers. *Proc Natl Acad Sci USA*, **109**(41), 16449–16454.

33. Kuo CK and Ma PX (2001). Ionically crosslinked alginate hydrogels as scaffolds for tissue engineering: Part 1. Structure, gelation rate and mechanical properties. *Biomaterials*, **22**, 511–521.

34. Thérien-Aubin H, Wu ZL, Nie Z and Kumacheva E (2013). Multiple shape transformations of composite hydrogel sheets. *J Am Chem Soc*, **135**(12), 4834–4839.

35. Jiang F, Huang T, He C, Brown HR and Wang H (2013). Interactions affecting the mechanical properties of macromolecular microsphere composite hydrogels. *J Phys Chem B*, **117**(43), 13679–13687.

36. Lu HH, El-Amin SF, Scott KD and Laurencin CT (2003). Three-dimensional, bioactive, biodegradable, polymer–bioactive glass composite scaffolds with improved mechanical properties support collagen synthesis and mineralization of human osteoblast-like cells in vitro. *J Biomed Mater Res*, **64A**, 465–474.

37. David R, Luu O, Damm EW, Wen JWH, Nagel M and Winklbauer R (2014). Tissue cohesion and the mechanics of cell rearrangement, *Development*, **141**, 3672-3682.

38. Cailliez F and Lavery R (2005). Cadherin mechanics and complexation: the importance of calcium binding. *Biophys J*, **89**, 3895–3903.

39. Korecky B, Hai CM and Rakusan K (1982). Functional capillary density in normal and transplanted rat hearts. *Can J Physiol Pharmacol*, **60**, 23–32.

40. Ott HC, Matthiesen TS, Goh SK, Black LD, Kren SM, et al. (2008). Perfusion decellularized matrix: using nature's platform to engineer a bioartificial heart. *Nat Med*, **14**(2), 213–221.

41. Inch WR, McCredie JA and Sutherland RM (1970). Growth of nodular carcinomas in rodents compared with multi-cell spheroids in tissue culture. *Growth*, **34**, 271–282.

42. Fennema E, Rivron N, Rouwkema J, van Blitterswijk C and de Boer J (2013). Spheroid culture as a tool for creating 3D complex tissues. *Trends Biotechnol*, **31**(2), 108–115.

43. Lou Y-R, Kanninen L, Kaehr B, Townson JL, Niklander J, Harjumäki R, Jeffrey Brinker C and Yliperttula M (2015). Silica bioreplication preserves three-dimensional spheroid structures of human pluripotent stem cells and HepG2 cells. *Sci Rep*, **5**, 13635.

44. Kaehr B, et al. (2012). Cellular complexity captured in durable silica biocomposites. *Proc Natl Acad Sci USA*, **109**, 17336–17341.

45. Meyer KC, Coker EN, Bolintineanu DS and Kaehr B (2014). Mechanically encoded cellular shapes for synthesis of anisotropic mesoporous particles. *J Am Chem Soc*, **136**, 13138–13141.

46. Townson JL, et al. (2014). Synthetic fossilization of soft biological tissues and their shape-preserving transformation into silica or electron-conductive replicas. *Nat Commun*, **5**, 5665.

47. Fisher OZ, Khademhosseini A, Langer R and Peppas NA (2010). Bioinspired materials for controlling stem cell fate. *Acc Chem Res*, **43**(3), 419–428.

48. Kloxin AM, Kloxin CJ, Bowman CN and Anseth KS (2010). Mechanical properties of cellularly responsive hydrogels and their experimental determination. *Adv Mater*, **22**(31), 3484–3494.

49. Slaughter BV, Khurshid SS, Fisher OZ, Khademhosseini A and Peppas NA (2009). Hydrogels in regenerative medicine. *Adv Mater*, **21**(32–33), 3307–3329.

50. Lou Y-R, Kanninen L, Kuisma T, Niklander J, Noon LA, Burks D, Urtti A and Yliperttula M (2014). The use of nanofibrillar cellulose hydrogel as a flexible three-dimensional model to culture human pluripotent stem cells. *Stem Cells Dev*, **23**(4), 380–392.

51. Gaharwar AK, Peppas NA and Khademhosseini A (2014). Nanocomposite hydrogels for biomedical applications. *Biotechnol Bioeng*, **111**(3), 441–452.

52. Gibson LJ and Ashby MF (eds) (1997). *Cellular Solids: Structure and Properties*. Cambridge University Press, Cambridge.

53. Karageorgiou V and Kaplan D (2005). Porosity of 3D biomaterial scaffolds and osteogenesis. *Biomaterials*, **26**, 5474–5491.

54. Woodard JR, Hilldore AJ, Lan SK, Park CJ, Morgan AW, Eurell JAC, Clark SG, Wheeler MB, Jamison RD and Johnson AJW (2007). The mechanical properties and osteoconductivity of hydroxyapatite bone scaffolds with multi-scale porosity. *Biomaterials*, **28**, 45–54.

55. Sánchez-Salcedo S, Arcos D and Vallet Regí M (2008). Upgrading calcium phosphate scaffolds for tissue engineering applications. *Key Eng Mater*, **377**, 19–42.

56. Janson IA and Putnam AJ (2015). Extracellular matrix elasticity and topography: material-based cues that affect cell function via conserved mechanisms. *J Biomed Mater Res Part A*, **103A**, 1246–1258.

57. Yang S, Leong KF, Du Z and Chua CK (2001). The design of scaffolds for use in tissue engineering. Part I. Traditional factors. *Tissue Eng*, **7**, 679–689.

58. Dutta RC, Dey M, Dutta AK and Basu B (2017). Competent processing techniques for scaffolds in tissue engineering. *Biotechnol Adv*, **35**(2), 240–250.

59. Venugopal J, Low S, Choon AT and Ramakrishna S (2008). Interaction of cells and nanofiber scaffolds in tissue engineering. *J Biomed Mater Res Part B: Appl Biomater*, **84B**, 34–48.

60. Seol YJ, Park JY, Jeong W, Kim TH, Kim SY and Cho DW (2015). Development of hybrid scaffolds using ceramic and hydrogel for articular cartilage tissue regeneration. *J Biomed Mater Res A*, **103**(4), 1404–1413.

61. Ryhanen J, Niemi E, Serlo W, Niemela KE, Sandvik P, Pernu H and Sato T (1997). Biocompatibility of nickel–titanium shape memory metal and its corrosion behavior in human cell cultures. *J Biomed Mater Res*, **35**, 451–457.

62. Li Z, Ramay HR, Hauch KD, Xiao D and Zhang M (2005). Chitosan-alginate hybrid scaffolds for bone tissue engineering. *Biomaterials*, **26**(18), 3919–3928.

63. Sheikh FA, Ju HW, Moon BM, Lee OJ, Kim JH, Park HJ, Kim DW, Kim DK, Jang JE, Khang G and Park CH (2016). Hybrid scaffolds based on PLGA and silk for bone tissue engineering. *J Tissue Eng Regen Med*, **10**(3), 209–221.

64. Wang D, Romer F, Connell L, Walter C, Saiz E, Yue S, Lee PD, McPhail DS, Hanna JV and Jones JR (2015). Highly flexible silica/chitosan hybrid scaffolds with oriented pores for tissue regeneration, *J. Mater. Chem. B*, **3**, 7560–7576.

65. Schagemann JC, Chung HW, Mrosek EH, Stone JJ, Fitzsimmons JS, O'Driscoll SW and Reinholz GG (2010). Poly-epsilon-caprolactone/gel hybrid scaffolds for cartilage tissue engineering. *J Biomed Mater Res A*, **93**(2), 454–463.

66. Jose MV, Thomas V, Johnson KT, Dean DR and Nyairo E (2009). Aligned PLGA/HA nanofibrous, nanocomposite scaffolds for bone tissue engineering. *Acta Biomater*, **5**, 305–315.

67. Gentile P, Chiono V, Carmagnola V and Hatton PV (2014). An overview of poly(lactic-co-glycolic) acid (PLGA)-based biomaterials for bone tissue engineering. *Int J Mol Sci*, **15**, 3640–3659.

68. Lee SB, Kim YH, Chong MS and Lee YM (2004). Preparation and characteristics of hybrid scaffolds composed of beta-chitin and collagen. *Biomaterials*, **25**(12), 2309–2317.

69. Gomes ME, Godinho JS, Tchalamov D, Cunha AM and Reis RL (2002). Alternative tissue engineering scaffolds based on starch: processing methodologies, morphology, degradation and mechanical properties. *Mater Sci Eng C*, **20**, 19–26.

70. Oh SH, Kang SG, Kim ES, Cho SH and Lee JH (2003). Fabrication and characterization of hydrophilic poly(lactic-*co*-glycolic acid)/poly(vinyl alcohol) blend cell scaffolds by melt-molding particulate-leaching method. *Biomaterials*, **24**(22), 4011–4021.

71. Gaudillere C and Serra JM (2016). Freeze-casting: fabrication of highly porous and hierarchical ceramic supports for energy applications. *Boletín de la Sociedad Española de Cerámica y Vidrio*, **55**(2), 45–54.

72. Hutmacher DW (2000). Scaffolds in tissue engineering bone and cartilage. *Biomaterials*, **21**, 2529–2543.

73. Mikos AG, et al. (1994). Preparation and characterization of poly (L-lactic acid) foam. *Polymer*, **35**, 1068–1077.

74. Gomes ME, Ribeiro AS, Malafaya PB, Reis RL and Cunha AM (2001). A new approach based on injection moulding to produce biodegradable starch-based polymeric scaffolds: morphology, mechanical and degradation behavior. *Biomaterials*, **22**, 883–889.

75. Monmaturapoj N, Soodsawang W and Thepsuwan W (2012). Porous hydroxyapatite scaffolds produced by the combination of the gel-casting and freeze drying techniques. *J Porous Mater*, **19**(4), 441–447.

76. Hsu YY, et al. (1997). Effect of polymer foam morphology and density on kinetics of in vitro controlled release of ionized from compressed foam matrices. *J Biomed Mater Sci*, **35**, 107–116.

77. Whang K, et al. (1995). A novel method to fabricate bioabsorbable scaffolds. *Polymer*, **36**, 837.

78. Smith IO, Liu XH, Smith LA and Ma PX (2009). Nano-structured polymer scaffolds for tissue engineering and regenerative medicine. *Wiley Interdiscip Rev Nanomed Nanobiotechnol*, **1**(2), 226–236.

79. Mooney DJ, et al. (1996). Novel approach to fabricate porous sponge of poly(D,L-lactic-co-glycolic acid) without the use of organic solvents. *Biomaterials*, **17**, 1417–1422.

80. Harris LD, Kim B-S and Mooney DJ (1998). Open pore biodegradable matrices formed with gas foaming. *Biomed Mater Res*, **42**, 396–402.

81. Mikes AG, Bao Y, Cima LG, Ingber DE, Vacanti JP and Langer R (1993). Preparation of poly(glycolic acid) bonded fiber structures for cell attachment and transplantation. *J Biomed Mater Res*, **27**, 183–189.

82. Freed LE, et al. (1994). Biodegradable polymer scaffolds for tissue engineering. *Biotechnology (NY)*, **12**, 689–693.

83. Gomes ME, Godinho JS, Tchalamov D, Cunha AM and Reis RL (2002). Alternative tissue engineering scaffolds based on starch: processing methodologies, morphology, degradation and mechanical properties. *Mater Sci Eng C Biomimetic Supramol Syst*, **20**, 19–26.

84. Jiang T, Abdel-Fattah WI and Laurencin CT (2006). In vitro evaluation of chitosan/poly(lactic acid-glycolic acid) sintered microsphere scaffolds for bone tissue engineering. *Biomaterials* **27**(28), 4894–4903.

85. Ruizab SA and Chen CS (2007). Microcontact printing: a tool to pattern. *Soft Matter*, **3**, 1–11.

86. Lackowski WM, Ghosh P and Crooks RM (1999). Micron-scale patterning of hyperbranched polymer films by micro-contact printing. *J Am Chem Soc*, **121**, 1419–1420.

87. Barbetta A, Rizzitelli G, Bedini R, Pecci R and Dentini M (2010). Porous gelatin hydrogels by gas-in-liquid foam templating. *Soft Matter*, **6**, 1785–1792.

88. Lee KY, Jeong L, Kang YO, Lee SJ and Park WH (2009). Electrospinning of polysaccharides for regenerative medicine. *Adv Drug Deliv Rev*, **61**, 1020–1032.

89. Kwon IK, Kidoaki S and Matsuda T (2005). Electrospunnano- to microfiber fabrics made of biodegradable copolyesters: structural characteristics, mechanical properties and cell adhesion potential. *Biomaterials*, **26**, 3929–3939.

90. Zhang Y, Ouyang H, Lim CT, Ramakrishna S and Huang ZM (2005). Electrospinning of gelatin fibers and gelatin/PCL composite fibrous scaffolds. *J Biomed Mater Res Part B: Appl Biomater*, **72**, 156–165.

91. Sun T, Mai S, Norton D, Haycock JW, Ryan AJ and Macneil S (2005). Self-organization of skin cells in three-dimensional electrospun polystyrene scaffolds. *Tissue Eng*, **11**(7–8), 1023–1033.

92. Zhang Y, Ouyang H, Lim CT, Ramakrishna S and Huang ZM (2004). Electrospinning of gelatin fibers and gelatin/PCL composite fibrous scaffolds. *J Biomed Mater Res Part B: Appl Biomater*, **72**, 156–165.

93. Kim HW, Kim HE and Knowles JC (2006). Production and potential of bioactive glass nanofibers as a next-generation biomaterial. *Adv Funct Mater*, **16**, 1529–1535.

94. Yeong W-Y, Chua C-K, Leong K-F and Chandrasekaran M (2004). Rapid prototyping in tissue engineering: challenges and potential. *Trends Biotechnol*, **22**(12), 643–652.

95. Hutmacher DW, Sittinger M and Risbud MV (2004). Scaffold-based tissue engineering: Rationale for computer-aided design and solid free-form fabrication systems. *Trends Biotechnol*, **22**, 354–362.

96. Wu BM, Borland SW, Giordano RA, Cima LG, Sachs EM and Cima MJ (1996). Solid free-form fabrication of drug delivery devices. *J Control Release*, **40**(1–2), 77–87.

97. Sobral JM, et al. (2011). Three-dimensional plotted scaffolds with controlled pore size gradients: effect of scaffold geometry on mechanical performance and cell seeding efficiency. *Acta Biomater*, **7**, 1009–1018.

98. Leong KF, Cheah CM and Chua CK (2003). Solid freeform fabrication of three-dimensional scaffolds for engineering replacement tissues and organs. *Biomaterials*, **24**, 2363–2378.

99. D'Urso PS, Earwaker WJ, Barker TM, Redmond MJ, Thompson RG, Effeney DJ and Tomlinson FH (2000). Custom cranioplasty using stereolithography and acrylic. *Br J Plast Surg*, **53**(3), 200–204.

100. Yilgor P, Sousa RA, Reis RL, Hasirci N and Hasirci V (2008). 3D plotted PCL scaffolds for stem cell based bone tissue engineering. *Macromol Symp*, **269**, 92–99.

101. Wu Y (2009). CAD-based interface programs for fusion neutron transport simulation. *Fusion Eng Des*, **84**(7–11), 1987–1992.

102. Berman B (2012). 3-D printing: the new industrial revolution. *Bus Horiz*, **55**(2), 155–162.

103. Rengier F, Mehndiratta A, von Tengg-Kobligk H, Zechmann CM, Unterhinninghofen R, Kauczor HU and Giesel FL (2010). 3D printing based on imaging data: review of medical applications. *Int J Comput Assist Radiol Surg*, **5**(4), 335–341.

104. Leukers B, Gülkan H, Irsen SH, Milz S, Tille C, Schieker M and Seitz H (2005). Hydroxyapatite scaffolds for bone tissue engineering made by 3D printing. *J Mater Sci Mater Med*, **16**(12), 1121–1124.

105. Woodfield TBG, et al. (2004). Design of porous scaffolds for cartilage tissue engineering using a three-dimensional fiber-deposition technique. *Biomaterials*, **25**, 4149–4161.

106. Zein I, Hutmacher DW, Tan KC and Teoh SH (2002). Fused deposition modeling of novel scaffold architectures for tissue engineering applications. *Biomaterials*, **23**, 1169–1185.

107. Hutmacher DW, Schantz T, Zein I, Ng KW, Teoh SH and Tan KC (2001). Mechanical properties and cell cultural response of polycaprolactone scaffolds designed and fabricated via fused deposition modeling. *J Biomed Mater Res*, **55**, 203–216.

108. Leong KF, Phua KK, Chua CK, Du ZH and Teo KO (2001). Fabrication of porous polymeric matrix drug delivery devices using the selective laser sintering technique. *Proc Inst Mech Eng H*, **215**, 191–201.

109. Paul BK and Baskaran S (1996). Issues in fabricating manufacturing tooling using powder-based additive freeform fabrication. *J Mater Process Technol*, **61**, 168–172.

110. Rimell JT and Marquis PM (2000). Selective laser sintering of ultra high molecular weight polyethylene for clinical applications. *J Biomed Mater Res*, **53**, 414–420.

111. Du Y, Liu H, Shuang J, Wang J, Ma J and Zhang S (2015). Microsphere-based selective laser sintering for building macroporous bone scaffolds with controlled microstructure and excellent biocompatibility. *Colloids Surf B*, **135**, 81–89.

112. Singh M, Sandhu B, Scurto A, Berkland C and Detamore MS (2010). Microsphere-based scaffolds for cartilage tissue engineering: using sub-critical CO_2 as a sintering agent. *Acta Biomater*, **6**(1), 137–143.

113. Bartl H and Von Bonin W (1962). About the polymerization in inverse emulsion. *Makromol Chem*, **57**, 74–95.

114. Bokhari M, Carnachan RJ, Przyborski SA and Cameron NR (2007). Emulsion-templated porous polymers as scaffolds for three dimen-

sional cell culture: effect of synthesis parameters on scaffold formation and homogeneity. *J Mater Chem*, **17**, 4088–4094.

115. Hayward AS, Sano N, Przyborski SA and Cameron NR (2013). Acrylic-acid-functionalized PolyHIPE scaffolds for use in 3D cell culture. *Macromol Rapid Commun*, **34**, 1844–1849.

116. Mi Y, Chan Y, Trau D, Huang P and Chen E (2006). Micromolding of PDMS scaffolds and microwells for tissue culture and cell patterning: a new method of microfabrication by the self-assembled micropatterns of diblock copolymer micelles. *Polymer*, **47**(14), 5124–5130.

117. Gallego D, Ferrell N, Sun Y and Hansford DJ (2008). Multilayer micromolding of degradable polymer tissue engineering scaffolds. *Mater Sci Eng C*, **28**(3), 353–358.

118. Koroleva A, Gill AA, Ortega I, Haycock JW, Schlie S, Gittard SD, Chichkov BN and Claeyssens F (2012). Two-photon polymerization-generated and micromolding-replicated 3D scaffolds for peripheral neural tissue engineering applications. *Biofabrication*, **4**, 025005 (11pp).

119. Nguyen KT and West JL (2002). Photopolymerizable hydrogels for tissue engineering applications. *Biomaterials*, **23**, 4307–4314.

120. Chu TMG, et al. (2001). Hydroxyapatite implants with designed internal architecture. *J Mater Sci Mater Med*, **12**, 471–478.

121. Hauser CAE and Zhang S (2010). Designer self-assembling peptide nanofiber biological materials. *Chem Soc Rev*, **39**, 2780–2790.

122. Gelain F, Horii A and Zhang S (2007). Designer self-assembling peptide scaffolds for 3-D tissue cell cultures and regenerative medicine. *Macromol Biosci*, **7**, 544–551.

123. Kokkoli E, Mardilovich A, Wedekind A, Rexeisen EL, Garg A and Craig JA (2006). *Soft Matter*, **2**, 1015.

124. Wu EC, Zhang S and Hauser CAE (2011). Self-assembling peptides as cell-interactive scaffolds. *Adv Funct Mater*, **22**, 456–468.

125. Matson JB and Stupp SI (2012). Self-assembling peptide scaffolds for regenerative medicine. *Chem Commun*, **48**, 26–33.

126. Nair AK, Gautieri A, Chang SW and Buehler MJ (2013). Molecular mechanics of mineralized collagen fibrils in bone. *Nat Commun*, **4**, 1–9.

127. Mann BK, Gobin AS, Tsai AT, Schmedlen RH and West JL (2001). Smooth muscle cell growth in photopolymerized hydrogels with cell adhesive and proteolytically degradable domains: synthetic ECM analogs for tissue engineering. *Biomaterials*, **22**(22), 3045–3051.

128. Yi JY, Yoon YH, Park HS, Kim CH, Kim CH, Kang HJ, et al. (2001). Reconstruction of basement membrane in skin equivalent; role oflaminin-1. *Arch Dermatol Res*, **293**, 356–362.

129. Weinberg CB and Bell E (1986). A blood vessel model constructed from collagen and cultured vascular cells. *Science*, **231**(4736), 397–400.

130. Fiegel HC, Kaufmann PM, Bruns H, Kluth D, Horch RE, Vacanti JP, et al. (2008). Hepatic tissue engineering: from transplantation to customized cell-based liver directed therapies from the laboratory. *J Cell Mol Med*, **12**(1), 56–66.

131. Mann BK, Tsai AT, Scott-Burden T and West JL (1999). Modification of surfaces with cell adhesion peptides alters extracellular matrix deposition. *Biomaterials*, **20**(23–24), 2281–2286.

132. Gobin AS and West JL (2003). Val–Ala–Pro–Gly, an elastin-derived non-integrin ligand: smooth muscle cell adhesion and specificity. *J Biomed Mater Res Part A*, **67A**(1), 255–259.

133. Griffith LG (2000). Polymeric biomaterials. *Acta Mater*, **48**(1), 263–277.

134. Salmon SE, Hamburger AW, Soehnlen B, Durie BG, Alberts DS and Moon TE (1978). Quantitation of differential sensitivity of human-tumor stem cells to anticancer drugs. *N Engl J Med*, **298**(24), 1321–1327.

135. Dermer GB (1994). Another anniversary for the war on cancer. *Biotechnology*, **12**, 320.

136. Hanahan D and Weinberg RA (2011). Hallmarks of cancer: the next generation. *Cell*, **144**, 646–674.

137. Nyga A, Neves J, Stamati K, Loizidou M, Emberton M and Cheema U (2016). The next level of 3D tumour models: immunocompetence. *Drug Discov Today*, **21**(9), 1421–1428.

138. Smith AST, Davis J, Lee G, Mack DL and Kim D-H (2016). Muscular dystrophy in a dish: engineered human skeletal muscle mimetics for disease modeling and drug discovery. *Drug Discov Today*, **21**(9), 1387–1398.

139. Govorkova EA, Murti G, Meignier B, de Taisne C and Webster RG (1996). African green monkey kidney (Vero) cells provide an alternative host cell system for influenza A and B viruses. *J Virol*, **70**, 5519–5524.

140. Wade N (2001). In tiny cells, glimpses of body's master plan. New York: New York Times, Dec. 18.

141. Kim BS and Mooney DJ (1998). Development of biocompatible synthetic extracellular matrices for tissue engineering. *Trends Biotech*, **16**, 224–230.

142. Polykandriotis E, Arkudas A, Horch RE, Stürzl M and Kneser U (2007). Autonomously vascularized cellular constructs in tissue engineering: opening a new perspective for biomedical science. *J Cell Mol Med*, **11**(1), 6–20.

143. Caspi O, Lesman A, Basevitch Y, Gepstein A, Arbel G, Habib IH, et al. (2007). Tissue engineering of vascularized cardiac muscle from human embryonic stem cells. *Circ Res*, **100**, 263–272.

144. Bhadriraju K and Chen CS (2002). Engineering cellular microenvironments to improve cell-based drug testing. *Drug Discov Today*, **7**, 612–620.

145. Galletti PM, Hellman KB and Nerem RM (1995). Tissue engineering: from basic science to products: a preface. *Tissue Eng*, **1**(2), 147–149.

146. Yannas IV and Burke JF (1980). Design of an artificial skin. I. Basic design principles. *J Biomed Mater Res*, **14**(1), 65–81.

147. Burke JF, Yannas IV, Quinby WC Jr, Bondoc CC and Jung WK (1981). Successful use of a physiologically acceptable artificial skin in the treatment of extensive burn injury. *Ann Surg*, **194**(4), 413–428.

148. Shu XZ, Ahmad S, Liu Y and Prestwich GD (2006). Synthesis and evaluation of injectable, in situ crosslinkable synthetic extracellular matrices for tissue engineering. *J Biomed Mater Res A*, **79**(4), 902–912.

149. Nguyen KT and West JL (2002). Photopolymerizable hydrogels for tissue engineering applications. *Biomaterials*, **23**(22), 4307–4314.

150. Kang SW, Jeon O and Kim BS (2005). Poly(lactic-co-glycolic acid) microspheres as an injectable scaffold for cartilage tissue engineering. *Tissue Eng*, **11**(3–4), 438–447.

151. Sung JH, Yu J, Luo D, Shuler ML and March JC (2011). Microscale 3D hydrogel scaffold for biomimetic gastrointestinal (GI) tract model. *Lab Chip*, **11**(3), 389–392.

152. Yu J, Peng S, Luo D and March JC (2012). In vitro 3D human small intestinal villous model for drug permeability determination. *Biotechnol Bioeng*, **109**(9), 2173–2178.

Chapter 4

Scaffolds/Matrices for 3D Cell/Tissue Culture

4.1 Introduction

As discussed in previous chapters, providing an in vivo like 3D microenvironment allows the cells to respond in a relatively natural manner. Even the physical mimicry of the native-like extracellular microenvironment/matrix (ECM) through organized three- dimensional physical support could bring the cell response closer to their in vivo like behavior [1]. Physical space in 3D permits them to expand in the third dimension as they would in the body. It also allows the cells to grow in close vicinity making the requisite cell-cell contact feasible. Such cell-cell bridges could happen only by chance or by virtue of the characteristics of a particular cell line in flat surfaces or 2D. For organotype culture this is essential and should happen naturally as the organized arrangement or cell-cell contact only leads to differentiation. Therefore, 3D scaffolds that mimic the physical attributes of natural ECM have proved useful though at times insufficient in achieving physiologically functional cell mass. Unless the 3D scaffold biomaterial is cell-interactive or cell-responsive, gathering or organizing physiologically active cells ex vivo is unexpected.

3D Cell Culture: Fundamentals and Applications in Tissue Engineering and Regenerative Medicine
Ranjna C. Dutta and Aroop K. Dutta
Copyright © 2018 Pan Stanford Publishing Pte. Ltd.
ISBN 978-981-4774-53-6 (Hardcover), 978-1-315-14682-9 (eBook)
www.panstanford.com

The acknowledgement of shortcomings in 2D is accompanied by the realization of the unavailability of a 3D system as the major obstacle in establishing new standards through organotypic cell culture [2]. We need an efficient cell-interactive scaffold of benign origin, preferably with good shelf-life and stability, to study cell-ECM dynamics in an experimental microenvironment [3]. Adapting organotype culture could reveal the true behavior of cells in real tissue like layout and also elaborate on the contextual cell-cell and cell-ECM dynamic relationship. Cell-ECM dynamics being at the helm of a fundamental understanding of normal vs. abnormal-cell response could thus provide an altogether new meaning to our approach towards therapeutics [4].

Numerous applications (Fig. 3.8) are possible with 3D scaffolds [5]. Nonetheless, the major impact of 3D cell culture is expected in drug development and diagnostics. Improved comprehension of bi-directional relation of the cell with its surrounding milieu has the potential to create a new line of drug design focused on empowering and restoring the natural microenvironment. This could create a new possibility of reverting an unhealthy or diseased condition. A well designed 3D scaffold may also be utilized to dispense and deliver therapeutic entities including cells [6]. However, the most ambitious purpose and goal of developing 3D scaffolds lies in engineering physiologically functional tissues so that the shortage of organs for transplantation could be dealt with. Scaffolds where technology allows host integration may be useful as a versatile therapeutic carrier to deliver pharmaceuticals as well as cells. They could also find a place in regenerative medicine if enabled to induce existing stem cells of the target tissue in a controlled manner (Fig. 4.1).

An ECM mimicking 3D scaffold could be of great help in accomplishing a physiologically functional tissue ex vivo. Figure 4.1 provides a schematic representation of possible applications of a 3D scaffold integrated with tissue-specific cues. A mimic of functional tissue ex vivo could establish a highly accurate model for basic research and drug discovery. Humanized tissue models would certainly be superior for diagnostics. They may even be evaluated and developed for studying metabolic pathways. However, in spite of well established role of ECM in the retention of physiological

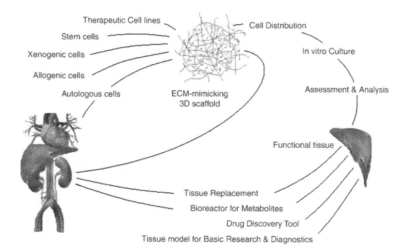

Figure 4.1 Potential of ECM mimicking scaffold in tissue engineering and regenerative medicine Ref. [4].

functions and tissue-like growth, only scanty literature is available that defines the significance of its individual constituents. It is likely that more than isolated conduct, the contribution of different bio-polymers on overall topography and the mechanistic aspect of ECM plays a role in bringing tissue specificity and resilience that keeps the cells/tissue healthy. Any modulation or shift in quality and/or quantity from natural healthy composition beyond the specified range is expected to have an adverse physiological impact. The topographical arrangement, very likely dependent on the relative composition of such ECM constituents is apparently crucial in channelizing any physiological activity too. Therefore, understanding the combination of ECM molecules and their stoichiometric ratio that brings tissue specificity might be vital. The decisive role of spatiotemporal interactions in bringing specificity at the molecular level has already been established through the structural study of isozymes and other receptor-ligand reactions that are capable of triggering a cascade of events [7]. Mechanisms of gene expression through transcription factors and the role of gene binding proteins in regulating genes also provide evidence in this respect.

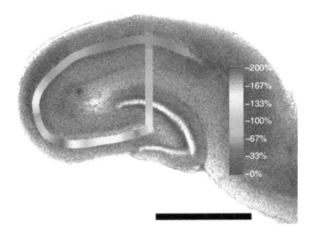

Figure 4.2   Elastic modulus variation (∼100 to ∼300 Pa/0% and 200% on the normalized color intensity scale bar) within the hippocampus of rat [reprinted from Elkin BS, Azeloglu EU, Costa KD and Morrison B. Mechanical heterogeneity of the rat hippocampus measured by atomic force microscope indentation. (Reprinted with permission, Copyright 2007 Mary Ann Liebert, Inc. Publishers)] (same picture is in Ref. [10]. As Fig. 2).

For engineering a specific tissue it is essential that the scaffold design, including its mechanics, should match the corresponding tissue. This is mainly because the matrix porosity and in turn its mechanical behavior is known to have considerable physiological implications on growing cells. The decisive influence of ECM stiffness on tissue functionality has already been established [8]. Likewise, the impact of matrix elasticity on the lineage pursued by the stem cells has also been reported [9]. As for the mammalian tissues, it is observed that the linear elastic modulus (LEM) or stiffness span over three orders of magnitude with each tissue falling in a certain specified range [10]. The variation in elastic modulus within a single tissue is also huge (Fig. 4.2). Table 4.1 below shows linear elastic modulus or stiffness of mammalian tissues largely measured by the compression method.

Understanding the level of molecular heterogeneity that exists in the target tissue ECM may prove helpful in providing primary guidelines for the design and fabrication of the scaffold.

Table 4.1 The linear elastic modulus or stiffness of mammalian tissues representing the range of the biomechanical nature of tissue and its microenvironment

Tissue	Stiffness/elastic modulus (Pa)	Reference
Fat	17	Wellman et al. 1999, Harvard Bio-Robotics Laboratory Technical Report.
Mammary gland	167	Paszek et al. 2005, *Cancer Cell.*
Brain	137–786	Elkin et al. 2007, *J Neurotrauma.* Gefen et al. 2003, *J Neurotrauma* (shear).
Liver	640	Yeh et al. 2002, *Ultrasound Med Biol.*
Kidney	7500	Nasseri et al. 2002, *Rheol Acta.*
Skeletal muscle	12000	Engler et al. 2004, *J Cell Biol*
Cartilage	949000	Freed et al. 1997, *PNAS.*
Bone	$4–400 \times 10^6$	Goldstein et al. 1983, *J Biomech* (shear).

A 3D scaffold optimized to match the native tissue ECM in terms of its constitution (fibrous, porous or combination), elasticity and compressive mechanical strength is thus a starting point for functional tissue modeling while cell interactivity, interconnected porosity, biocompatibility and controlled degradation for swapping with neo-ECM remain other essential features. Cell response is also influenced by the size and extent and sometimes the hierarchical arrangement of the porosity of the 3D scaffold [11, 12]. This also needs to be optimized for different tissues. Architectural similarity with native ECM could possibly be generated through gradient features within a certain specified range [13].

Numerous scaffolds constituted from a varying mix of homo, hetero or co-polymers through different fabrication techniques have been explored for organoid culture [14]. Polymers that form hydrogels are found particularly useful as they could retain sufficient water and grossly match with natural ECM in hydration characteristics [11, 15]. The appropriateness of biomaterial depends not only on the tissue to be engineered but also the technique utilized in creating the scaffold [16]. Techniques involving harsh conditions are not suitable for processing natural polymers as they may alter them adversely or lead to denaturation. Collagen, fibrin, gelatin, albumin, cellulose, sodium hyaluronate, silk, and alginate are some of the commonly used natural polymers while polyvinyl alcohol, polyethylene glycol, poly N-isopropylacrylamide, and poly-(lactic acid-glycolic acid) copolymers are the most customary synthetic polymers tested for the purpose [17, 18]. Hydroxyapatites and tricalcium phosphates like bio-ceramics are also recommended especially for engineering artificial bone and cartilage [19]. However, both synthetic and naturally derived biopolymers have their own limitations. Natural biomaterials are cell-interactive but delicate to handle, difficult to reproduce and limited by source. Sterility and immunological issues may also be associated with them. Synthetic materials on the other hand are well defined, reproducible and sturdy but unable to impart cell interactivity and tissue specific cues beyond the physical framework. Most often they also exhibit unacceptable degradation kinetics and byproducts. Biomaterial and the resultant scaffold should also be immunologically inert if they are to be used for in vivo purposes. This has encouraged the exploration of some hybrid biomaterials also for the purpose.

The three-dimensional structure of a scaffold can be described effectively based on its size, shape and surface texture. It has been postulated that the nanotopography of the scaffold plays a major role and influences cell orientation and migration. It was also demonstrated recently that the orientation and migration of osteoblast which is a prerequisite for directed bone tissue formation is significantly enhanced on nanostructured surfaces [20]. The development of new biomaterials, technologies and scaffolds like Extracell®, ECM-Analog®, BD-Matrigel®, Corning-Matrigel®, Alvetex®, BioVaSc-TERM®, Algimatrix® and spheroids of 3D-Biotek® that might facilitate organotypic culture and their commercial availability, especially in conventional plate formats, are expected to bring the desired progression in our knowledge. Given some efforts, a few of these technologies also have the potential to graduate/advance to in vivo applications.

4.2 Non-specific in vitro 3D Culture

Technologies that are based on synthetic polymers or involve major synthetic modifications of natural polymers could be used only for in vitro 3D culture purpose. A natural biomaterial with significant modification, though enabled to sustain living cells in vitro, needs to be assessed thoroughly for its safety before qualifying for in vivo applications. Synthetic alterations of natural biomaterial should have a minimal impact on their biocompatibility; else they may lead to unpredictable or even adverse implications if used in vivo.

Conventional methods of cell culture has established the benign role of polystyrene like inert synthetic polymers on cells. This led to the creation of 3D fibrous network of polystyrene used in Alvetex or 3D-Biotek culture wares. Polystyrene fibers are used to provide 3D space to the growing cells. However, they cannot be used for in vivo purposes. Also, owing to the inert nature of the scaffold, compromised and/or deviated cell-cell and cell-ECM interactions are anticipated. Cells grown in such systems may be better than 2D but are not expected to represent their true physiology.

4.3 Tissue Engineering

The need for engineered tissue implants with an adequate biological approach was realized during the period of World War [21], although the term "tissue engineering" was coined only in 1987 in the National Science Foundation meeting. This was followed by the first tissue engineering workshop at Lake Tahoe in 1988 [22]. Tissue engineering (TE)/regenerative medicine (RM) research truly accelerated in the 1990s and successes in the field culminated into a blooming industry in a short span of time. Tissue-engineered skin substitutes are the earliest clinically proven products but in the absence of an interdisciplinary environment, poor understanding of regulatory, financial and other product specific needs, they are yet to be translated to commercial success [23].

It is pertinent to emphasize that engineering a tissue-equivalent is not possible without 3D cell culture. Cells in nature co-exist and co-ordinate to impart a unique functionality manifested in the form of a tissue. Growing and organizing cells in vitro in the form of tissue as cohesive and dense as their counterparts in vivo without losing their functionality had been a huge challenge. Replicating tissue specific microenvironment artificially is unimaginable unless we specify the explicit role of at least the major ECM components. Knowing the source and factors that regulate the tissue specific microenvironment will certainly provide us with the much needed control over the functionality of the artificially growing cellular mass. Only 3D cell culture enabling technologies promote tissue-like growth and this comprehension is the basis of extensive research in biomaterials and scaffolds. A synchronized effort to develop tissue specific ECM mimicking scaffolds and processes for isolating and cultivating relevant cells ex vivo led TE technologies to its present status. It is now clear that cell-interactive 3D scaffolds designed and processed with appropriate biomaterial for yielding optimal porosity and mechanical strength is the key to engineer functional tissue ex vivo. In order to accomplish these facets in a 3D scaffold, multiple techniques and processes have been explored by various research groups all over the world (refer to Chapter 3).

Hydrogels and scaffold matrices created from natural biopolymers having safe degradation kinetics and products in vivo are the most suitable technologies for TE and RM. However, being delicate to handle and often difficult to hold cells within, not all the hydrogels derived from natural biomaterials are suitable for the purpose. Hydrogels of natural biopolymers therefore need crosslinking or an additional polymer to strengthen the scaffold. Integra®, Primatrix® and ECM Analog® are the only few technologies that offer ex vivo tissue engineering. BioVaSc-TERM may also be used in certain instances especially where pre-existing vascularization would help, though it may require pre-determination of immuno-compatibility with the host. Collagen derived scaffolds in certain cases are found very promising due to the fact that the polymer is a major native constituent of the mammalian extracellular matrix. It has been revealed during digestive studies that collagen proteins get morphological features through their relative association with different carbohydrate moieties. Not only their ratio and quality but also their interstitial location helps in modulating the mechanical features of collagen. Apart from glucose and galactose, collagens may also contain small amounts of other sugar components including mannose, fucose, hexosamines, and sialic acids [24].

4.4 Regenerative Medicine

Regenerative medicine (RM) involves repairing a tissue that has been partially damaged or dead. Tissue regeneration or repair may be accomplished through the following two approaches:

4.4.1 In situ

In situ approach involves molecules capable of launching the repair mechanism by inducing the localized cells on-site in a target specific manner. This can be achieved with or without carrier matrices or hydrogels that could be administered even as injectables. In situ tissue regeneration may also be achieved through a drug molecule or an entity that can instigate the stem cells to commit to a desired differentiation path by modulating their niche. However, this approach remains in the distant future as the research and understanding about differentiation, de-differentiation and trans-differentiation is still in its infancy. Recent research has identified cancerous stem cells, putting forth another caution and challenge in following this approach. It is extremely important to identify and kick off the right kind of cells from the stem cell cluster present in the niche to be successful in this approach.

4.4.2 Ex situ

Ex situ approach involves the use of scaffolds or matrices capable of integrating with the damaged or dysfunctional or nonfunctional tissue. It requires growing functionally viable cells ex vivo before they are ready for implantation. For better integration with the host tissue this approach demands matrix and scaffolds with tissue specific mimicking. Unlike the in situ approach, this requires some ex vivo culture period for yielding functionally viable cells in sufficient numbers before implantation.

4.5 Available Technologies

Multiple research groups all over the world are working on the existing challenges in tissue engineering and regenerative medicine. While companies have developed different technologies to create 3D scaffolds through engineering principles, the product formats are still evolving. Many of the products appear and disappear from their websites after sometime. The reasons for such a short appearance may vary from product shelf-life, application potential or demand, the simplicity of handling and the urgency of that kind of product. Awareness, affordability or competitive cost are other important factors to gauge their popularity. Some of these products have evolved over time but others remain transient or have become redundant.

There has been a surge of interest in biomaterial scaffold development for tissue engineering and, for that matter, in drug delivery space over the last two decades. This has resulted in a burst of new startups, especially in the United States, a country known for its adventurous and supportive environment for new startups. Financial support from the National Institute of Health (NIH) along with generous company policies and technical backup from well equipped university infrastructure creates a conducive environment for a new startup to grow and remain industrious and motivated. Europe and Asia are still working and evolving their startup policies to match the rate of proliferation as happens in the US. Table 4.2 includes some of the technologies in use for creating scaffolds/matrices and indeed majority of them are developed in the US. The list is not all inclusive but most of the scaffolds/matrices are being offered without cells; leaving the application options open for the users. Few products are also being explored with cells already incorporated within the matrices. This is attempted especially with hydrogels which also hold promise as injectables and could be used as carrier for the cells. Some of the scaffolds could be used as an implant with or without cells. Few of them are also suitable for autologous and stem cell culture and therefore have great potential in organotypic culture and/or engineering tissues ex vivo (Table 4.2).

Table 4.2 Commercially available technologies/scaffolds, their source/constituents and feasibility for in vitro 3D cell-culture and/or in vivo TE and RM applications

S. No.	Technology/scaffold	In vitro 3D culture	Organotype/ ex vivo TE	Regenerative	Origin (reasons for applicability)
1.	Matrigel	Yes	No	No	Tumor origin
2.	Alvetex® (Re-innervate)	Yes	No	No	Polystyrene
3.	3D-Biotek®	Yes	No	No	"
4.	AlgiMatrix®	Yes	Yes	Yes	Alginate
5.	PuraMatrix®	Yes	No	Yes	16-mer peptide
6.	Integra®	Yes	Yes	Yes	Bovine tendon collagen
7.	PriMatrix	Yes	Yes	Yes	Fetal bovine dermis (Col-III dominant)
8.	Hyalubrix® /hyalgan	Yes	No	Yes	Hyaluronic acid
8.	Extracel™	Yes	Yes	No	Thiol-modified hyaluronan & PEGDA
9.	Mebiol	Yes	No	No	Copolymer of poly-(N-isopropylacrylamide and PEG
10.	UpCell™	Yes	No	No	poly-(N-isopropylacrylamide)
11.	BioVaSc-TERM®	Yes	No	No	Decellularized Porcine jejunum
12.	Corgel®	Yes	No	No	Tyramine-modified Hyaluronan
13.	Oasis® Wound Matrix	Yes	Yes	No	Porcine small intestinal submucosa (SIS)
14.	Cyto-Matrix	No	No	?	Herbal formulations
15.	Amniograft	No	No	Yes	Human amniotic membrane
16	Artiss/Tisseel	No	No	Yes	Fibrin matrix as glue
17.	ECM Analog®	Yes	Yes	Yes	Hydrolyzed/denatured collagen

6 hours   16 hours   24 hours

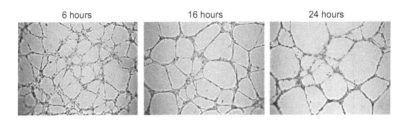

Figure 4.3 Human umbilical vein endothelial cell (HUVEC) tube formation on a basement membrane substrate with time. Cells were seeded at 4,800 cells/cm^2 (part of Fig. 2 in Arnaoutova et al., *Angiogenesis*, 2009, **12**, 267–274.

4.5.1 Matrigel

Matrigel is a gel-like matrix derived from the Engelbreth-Holm-Swarm (EHS) mouse sarcoma. Sarcoma, a type of cancer, is associated with mesenchymal cells. Mesenchymal cells represent typical non-polar cells surrounded by a plentiful extracellular matrix. These cells are the basic constituents of connective tissues like the lymphatic and circulatory system and also of bone and cartilage.

Matrigel was first used for establishing the influence of 3D space. Cells responded more like their in vivo counterparts when grown in matrigel [25]. As a scaffold matrigel exhibits peculiar rheological properties, it remains a liquid at 4 degrees and forms a gel at 37 degrees (body temperature). Human umbilical vein endothelial cells (HUVEC) exhibit tube formation when grown on Matrigel (Fig. 4.3). As a basement membrane substrate, Matrigel allows cells to grow in 3 dimensions; however, being of cancerous origin it has limited value in the applications like TE and RM. In the absence of better alternatives, matrigel is frequently chosen for in vitro 3D cell culture [26, 27], studying tumor cell metastasis and cancer drug screening [28].

Matrigel is now commercially available. It is produced and marketed by Corning Life Sciences and BD Biosciences. Trevigen Inc. has started marketing one of its own versions of Matrigel under the trade name Cultrex BME. It remains a benchmark till today and is frequently used as a control for comparison purposes.

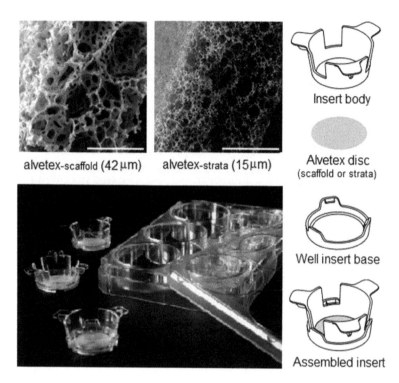

alvetex-scaffold (42 µm)   alvetex-strata (15µm)

Insert body

Alvetex disc
(scaffold or strata)

Well insert base

Assembled insert

Figure 4.4 Integra products manufactured using bovine tendon collagen and glycosaminoglycan on silica bilayer.

4.5.2 Alvetex®

Alvetex® is a highly porous polystyrene scaffold designed for 3D cell culture. As a 200 micron thick membrane Alvetex is adapted to fit into conventional cell culture plastic wares. Earlier called reinnervate (www.reinnervate.com), it is supplied by ReproCELL Europe Ltd. sterilized through gamma irradiation, the Alvetex-scaffold is recommended to be washed with ethanol prior to use. There are two basic formats: Alvetex® scaffold and Alvetex® strata that differ in the average void size, that is 42 µm in scaffold and 15 µm in strata (Fig. 4.4).

Matrices in both the Alvetex-scaffold and Alvetex-strata are presented as 200 micron thick porous cross-linked polystyrene where the only difference is in their architecture. Both types of membranes are also available as inserts for use in multiwell culture plates. However, since the constructing material is polystyrene there is not much difference from conventional 2D culture plates in cell seeding and handling. This makes it convenient and close to conventional methods. Besides, little difference is expected from the cell response point of view [29].

4.5.3 3D-Biotek®

3D-Biotek® (http://www.3dbiotek.com/) is a US-based cell culture-ware company that also uses polystyrene to create scaffolds as tools for growing cells in 3 dimensions. The scaffolds and inserts are manufactured using precision microfabrication technology. They opened the market with 3D inserts™ and are now developing culture wares like 3DKUBE™ perfusion bioreactors, etc., suitable for 3D culture in vitro. Their scaffold is meant for generating tumor spheroids for drug screening, especially in cancer biology (http://www.3dbiotek.com/documents/Anchorage_Spheroids.pdf).

By providing physical space in 3D even through plastic, such tools are suitable to take the existing flat surface cell-culture practice to the next level. Cells are expected to grow and respond better compared to 2D platforms yet likely to remain distant from the native in vivo like response. This would be mainly because the material used is not cell interactive and therefore insufficient to provide an ECM-like microenvironment to growing cells.

4.5.4 AlgiMatrix® 3D Cell Culture System

AlgiMatrix® is a lyophilized sponge created from alginate, a polysaccharide extracted from seaweed. Its macroporous structure allows easy cell loading. AlgiMatrix being made up of biodegradable well-defined biomaterial is usable for organotypic 3D cell culture. It can also be used for creating spheroids and therefore for assays that involve multicellular tumor spheroids [30]. For cell-based screening and tests in the 3D format, AlgiMatrix is already being marketed through Thermo Fisher Scientific (https://www.thermofisher.com) as a suitable substrate. Apart from scaffolds, it is also available in multiwell plate formats.

The unique feature of the alginate matrix is easy control of its material properties. Polymerization and gelling of alginate is sensitized through calcium ions. So the firmness of the matrix can be controlled to a great extent simply by changing the calcium concentration in the medium. AlgiMatrix® is particularly found useful in the field of liver biology. Hepatocellular spheroids which are otherwise cumbersome to keep in culture are shown to maintain the expression of key hepatic markers when grown on AlgiMatrix.

It has been demonstrated that the AlgiMatrix environment is less stressful to hepatocytes in comparison to conventional sandwich culture. It has also been observed that when grown on AlgiMatrix, primary hepatocytes respond specifically and quantitatively to model CYP450 inducers. This makes it a preferred tool for predicting xenobiotic metabolism and toxicology research.

4.5.5 PuraMatrix®

PuraMatrix® (http://www.puramatrix.com/) is the only technology similar to the natural hierarchical assembly of molecules that form the hydrogelmatrix through electrostatic and hydrogen bonding. The basic unit of PuraMatrix is a 16-mer peptide that includes repeating amino acid sequences of Arginine-Alanine-Aspartic acid-Alanine (Fig. 4.5). When exposed to physiological levels of salt, PuraMatrix strands self-assemble to form nanofibrous, porous scaffolds in 3D. The nanofiber density and average pore size (5–200 nm) can be controlled and customized by the concentration of peptide [31].

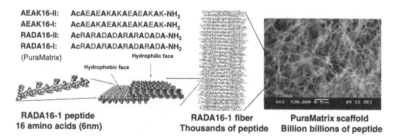

Figure 4.5 Hierarchical assembly of PuraMatrix (adapted from Fig. 1a of Ref. [31]).

PuraMatrix® has been commercially available for research applications since 2004 via Becton Dickinson and is currently sold by Corning, Inc. It is also available with the CE mark as a clinical, surgical hemostatic agent under the trade name PuraStat®.

Owing to its simplicity and ease of handling, PuraMatrix has proven its utility in different areas of research before being declared as a commercial product (http://puramatrix.com/wp/publications). Hierarchical versatility similar to the natural system has allowed morphological control over a wider scale. Furthermore, the modulation of gelling properties simply through concentration variations has made it possible to use it as an injectable material. It has been evaluated alone and in combination with other biomaterials for the purposes of hemostasis, bone-cartilage regeneration, cardiovascular, neurosurgical and soft tissue regeneration related studies.

4.5.6 Integra®

Integra® represents a porous matrix of crosslinked bovine tendon collagen and glycosaminoglycan on a silicone bilayer. The biodegradable matrix provides a scaffold for cellular invasion and capillary growth. The semipermeable polysiloxane or silicone layer helps in controlling water vapor loss while providing a flexible adherent covering for the wound surface. This also provides strength to the matrix and facilitates handling. The Integra matrix is now available under Integra Life Sciences (www.integralife.com) mainly in sheet format, which is meant for different indications. Products vary either in thickness or in constituting biomaterial. Their use is suggested for different indications based on crosslinking, incorporation of the allogenic placental matrix, and thickness of the silicon layer sheet used for spreading the matrix and granulations.

(i) *Integra® bilayer wound matrix*: It is meant as a soft tissue solution for the lower and upper extremities in burn and wound management. It is also applicable for plastic and reconstructive surgery. It is expected to manage partial and full thickness wounds, pressure and venous ulcers, chronic and vascular ulcers and also diabetic ulcers. It can also tend surgical wounds both at the donor or graft sites. Trauma wounds including abrasions, lacerations, skin tear and second degree burns and draining wounds can also be taken care of by using this matrix device.

(ii) *Integra® BioFix® flow placental tissue matrix allograft*: This is an allograft which involves human placental tissue. The proprietary processing technology used for the product is supposed to preserve the natural structure and biological properties of the tissue. Derived from the decellularized particulate human placental connective tissue matrix, it represents the allogenic ECM developed with the intention of a broad range of clinical applications. It is supplied in different volumes as a free flowing sterile allograft and is also suggested for managing the lower and upper extremities, and burns and wounds.

(iii) *Integra® flowable wound matrix*: The matrix in this case is in granular form that could be hydrated with saline and applied to uneven and diverse wound sites. The flowable wound matrix could also be administered through syringes or injectors, depending on the crosslinking of the granular matrix. Flexibility of administration allows its use in multiple indications, including the lower and upper extremities, and burn and wound management.

(iv) *Integra® HuMend™ acellular dermal matrix*: This product acts as a supportive reinforcing scaffold in soft tissue repair and tendon protection. It is created with the intent to use it for abdominal wall and breast reconstruction.

(v) *Integra® meshed bilayer wound matrix*: This has been designed for negative pressure wound therapy where the bilayer, made up of a porous matrix over semipermeable polysiloxane, allows drainage of wound exudate.

(vi) *Integra® reinforcement matrix*: Here the Integra matrix from acellular porcine dermal matrix is optimized for tensile strength to support and reinforce the soft tissues repaired by sutures or suture anchors during tendon repair. This helps in preventing tearing from the sutured site.

(vii) *Integra® wound matrix and Integra® wound matrix (thin)*: The Integra™ wound matrix is designed to provide coverage over exposed bone, tendon, cartilage and joints. They are similar to the basic matrix except they posses 50% less collagen compared to each of the corresponding square centimeter sizes of the normal wound matrix.

4.5.7 PriMatrix

The PriMatrix® dermal repair scaffold is created by using fetal bovine dermis for the management of wounds. An acellular dermal tissue matrix, PriMatrix is rich in type III collagen, which is said to be active in developing and healing tissues. It is made available in different sizes in meshed, fenestrated and solid configurations by Integra® only. It is suggested for the covering/management of skin ulcers, second degree burns and trauma and surgical wounds of partial and full thickness types [32].

PriMatrix promises to possess the dermal tissue matrix, including the naturally linked collagen fibers that allow easy cellular interactions with the host site. However, being rich in collagen of bovine origin it cannot be recommended to patients with a known history of hypersensitivity to collagen or bovine products (www.primatrix.com). The matrix shrinks if hydrated at a little higher temperature. This may also lead to complications like chronic inflammation, allergic reactions, excessive redness, pain, swelling or blistering in certain subjects. It may also cause infection or add to existing infection if not handled aseptically.

4.5.8 Hyalubrix®/Hyalgan

Hyalubrix® is a sterile, non-pyrogenic, viscoelastic solution prepared from the high molecular weight (>1500 kDa) fraction of hyaluronic acid sodium salt produced through bacterial fermentation. The product is available in the form of a prefilled syringe for intra articular injections. Hyaluronic acid (HA) that belongs to a glycosaminoglycan family of polysaccharide is naturally present in the cartilage and synovial fluid. It is believed that HA is the major contributor to the viscosity and elasticity of the synovial fluid. The lubricating and shock absorbing functions exerted by the fluid in normal joints is also attributed to HA. Lubrication actually protects cartilage and soft tissue on the joints against mechanical injuries. Loss of lubrication due to changes in viscosity or deficiency of HA in the joints is considered a major morbidity in traumatic and degenerative joint disorders. Stiffness of joints as in arthritis leads to painful and/or impairment of joint functions The intra-articular administration of HA is capable of restoring the viscoelastic properties of synovial fluid and helps in reducing the pain and improving joint mobility.

Hyalubrix® is available in a syringe prefilled with a 2 ml sterilized solution of 30 mg HA sodium salt. Each filled syringe is then sealed in a blister pack sterilized by ethylene oxide. A comparative double blind, controlled study of Hyalubrix® injections versus a local anesthetic in osteoarthritis of the hip is found to be safe and have beneficial effects [33].

Hyalubrix® (http://www.hyalubrix.it/) is also marketed by Fidia Farmaceutici (http://www.iafstore.com/ita/fidia-farmaceutici/hyalubrix-codp22615).

4.5.9 Extracel™

Extracel™ hydrogel is a hyaluronan-gelatin-based hydrogel. The hyaluronan-based product is also mentioned as Extracel-HP™ hydrogel at some places. Extracel represents a semi-synthetic matrix as it uses thiol-modified hyaluronan, thiol-modified gelatin (denatured collagen) and a thiol-reactive crosslinker polyethylene glycol diacrylate (PEGDA). This hydrogel is stated to be optimal for culturing primary cells and cell lines in 3D.

It is marketed by Glycosan Biosystems Inc. (www.glycosan.com), Advanced BioMatrix (www.AdvancedBioMatrix.com) and also by CellSystems biotechnologie (www.cellsystems.de). Operating as a subsidiary of OrthoCyte corporation, Glycosan Biosystems Inc. offers Extracel and other hyaluronan-based products like HyStem-C as a starting point for optimization of the stem cell's microenvironment; Extracel-HP and HyStem-HP for the slow release of growth factors in the cell's microenvironment; Extracel-X for growing tumor xenografts in mice; Extracel-SS and HyStem-CSS that allows recovery of encapsulated cells in 3D cultures.

4.5.10 Mebiol

Mebiol is thermo-sensitive and exhibits reversible, temperature-based sol-gel transition. The copolymer poly(N-isopropylacrylamide) and poly(ethylene glycol) that constitutes Mebiol has the unique property of remaining in a liquified or sol state when chilled and converts into a solid hydrogel format when heated. It is a thermoreversible hydrogel of synthetic polymers with high transparency. Stem cell and pluripotent stem cell culture, expansion and differentiation, spheroid culture, cell implantation, organ and tissue regeneration, drug delivery, etc., are some of the possible applications suggested for Mebiol. Although Mebiol suppresses fibroblast growth, it has been commercialized as a cell/tissue culture reagent for embryonic stem cells, chondrocytes and cancer cells. It is also marketed by CellSystems.

Cell systems' website also displays availability of reconstructed human epidermis (RhE) using Extracel technology. It was initially named as "Epidermal Skin Test-EST-1000®", later "EST-1000®" and then in 2012 it is named as "epiCS®". Cell systems have also launched the pigmented RhE-"epiCS®-M" which is available with melanocytes from donors of three ethnic groups: Caucasian, Afro-American and Asian-Caucasian.

The reconstructed human epidermis (RhE) model epiCS developed at Cell Systems is a 3D model of human epidermis. It is a multi cell layer model derived from normal human keratinocytes that mimics the biochemical and physiological properties of the epidermis. It is, therefore, considered as a human epidermis equivalent for in vitro studies. The cellular structure of RhE closely resembles the epidermis, basement membrane, proliferating keratinocytes and stratum corneum with a barrier function. This allows topical application of liquid, creamy and also solid substances on the model. It can also be used for studying systemic exposure of different molecules by adding them to the cell culture medium. These models are now validated and accepted by regulatory authorities like the European Centre for the Validation of Alternative Methods (ECVAM) as a stand-alone method for skin corrosion testing and in vitro toxicology studies.

4.5.11 UpCell™

UpCell™ also utilizes the thermo-sensitive synthetic polymer hydrogel as the core technology in providing cells an ambience better than 2D. It is expected that the cell culture plates and wells coated with a thermo-responsive poly-(N-isopropylacrylamide) or PNIPAAm would support the growing cells and also allow them to be retrieved non-enzymatically. The cells are to be seeded in a sol state (at room temperature) and put in an incubator where the coating gels solidify while allowing the cells to adhere and proliferate. After confluence, the cell sheet can be transferred through the supportive membrane from the polymer gel to another substrate for imaging and analysis by lowering the temperature from 37 to 32 degrees. Nunc™ has adopted UpCell™ technology for its cell culture dishes, microplates and flasks and marketing them. It is available as Nunc™ Dishes with UpCell™ Surface. UpCell® is also being promoted by CellSeed, a Japanese company. RepCell® and HydroCell™ are other two products based on the technology. Ease of handling cells that reach to confluence is certainly the USP of this technology. Often referred to as cell sheet engineering, it is proposed as an additive technology. It allows cells to grow in sheet format presumably through a self-generated extracellular matrix.

The convenience of handling confluent cells that might stick together by virtue of their native characteristics is driving the use of the technology. However, the toxicity related to acrylates even as remnants or leached out in traces around the cells at different stages of growth and its impact post-transplantation cannot be ignored while exploring their application in tissue engineering and regenerative medicine. Nonetheless, it is suggested that the cell sheets grown in vitro could be transplanted to host tissues without the need of any supporting scaffold or carrier material. Tissue constructs using stacks of multiple individual cell sheets created in vitro are also proposed for endoscopic transplantation, combinatorial tissue reconstruction, and polysurgery to overcome limitations of regenerative therapies and cell delivery with conventional approaches [34]. Detailed pre-clinical evaluation is required for validating these assumptions.

4.5.12 BioVaSc-TERM®

BioVaSc-TERM represents a biological vascularized scaffold platform that is based on porcine jejunal segments which are decellularized and sterilized (www.biovascularinc.com and www.lifesciences.fraunhofer.de). The vascular structures from the arterial pedicles and venous returns to the fine capillary structures are preserved and can be reseeded with human endothelial cells [35]. This scaffold can be connected to a fluidic system in customized bioreactors providing appropriate dynamic culture conditions. To generate customized tissue models, further cell types can be included. Moreover, the BioVaSc-TERM® can be connected to a patient's bloodstream after implantation, representing a unique feature of this technology.

This scaffold is being successfully utilized for the reconstruction of extensive airway defects [36]. Using the BioVaSc-TERM complex, vascularized human full skin model (SkinVaSc-TERM®) is generated [37]. The technology has a potential to be extended to other tissues like Gut (GutVaSc-TERM®), Lungs (LunVaSc-TERM®), Oncology (OncoVaSc-TERM®), Pancreas (PanVaSc-TERM®), Stomach (StoVaSc-TERM®) and Trachea (TraVaSc-TERM®).

4.5.13 Corgel®

Corgel BioHydrogel is a biocompatible, hyaluronan hydrogel marketed by Lifecore (http://www.lifecore.com). Lifecore is an established manufacturer and supplier of pharmaceutical and medical grade sodium hyaluronate. For commercial purposes, sodium hyaluronate is produced by microbial fermentation and purified.

The hydrogel is based on crosslinked tyramine substituted sodium hyaluronate (TS-NaHy) [38]. The modulation of physical properties of the hydrogel is possible through the percentage of tyramine substitution and also its concentration in the formulation [39]. The hydrogels formed using TS-NaHy solutions in the range of 6.25 to 100 mg/ml display a spectrum of physical and rheological attributes. They culminate into a weak gel, paste or a fractile solid, which is made available as 1, 3, and 5% tyramine substituted hyaluronan. Low level substitution and crosslinking maintains the native structure of the sodium hyaluronate.

Through a pilot study in a canine model Fumoto et al. have shown the feasibility of treating mitral regurgitation through tyramine-based hyaluronan hydrogel injection into the base of the left ventricle between the two papillary muscles [40]. Mitral regurgitation is a condition where the blood leaks backward through the mitral valve each time the left ventricle contracts. Leakage can increase blood volume and pressure in the left atrium leading to increasing pressure in the pulmonary veins (veins that connect lungs to the heart). If severe, increased pressure may result in congestion (or fluid build-up) in the lungs and other complications. Severe regurgitation may cause palpitation, enlargement of heart and even heart failures.

4.5.14 Opsite, Biobrane and Oasis® Wound Matrix

Oasis® matrix is a porcine derived, intact 3-dimensional extra-cellular matrix (ECM) scaffold marketed by Smith and Nephew. Oasis® Wound Matrix and Oasis® Ultra Tri-layer Matrix represent a single and 3 layers of porcine small intestinal submucosa (SIS). It is indicated for the management of all types of wounds and ulcers. That it helps in handling partial and full-thickness wounds of surgery (donor sites/grafts, post laser, etc.), trauma (abrasions, lacerations, skin tear, second degree burn, etc.) and tunnel type and pressure and diabetic ulcers is also suggested.

However, it is recommended only after excessive bleeding, exudate, swelling and infection is controlled to avoid complications like allergic reactions, inflammation, redness, pain, blistering or redness. It should not be used in patients with known sensitivity to porcine material [41]. Use of the Oasis® wound matrix is therefore suggested under strict clinical supervision.

The range of OPSITE films involves acrylic adhesive that makes the film extensible and also waterproof. The film also helps in the adequate retention of natural wound exudate.

Opsite Post-op is another range of dressings made up of SIS but designed for post-operative visibility and examination without removal. With a highly absorbent lattice pad covered with a transparent bacteria-proof barrier on the top and a conformable acrylic sheet below, it manages the exudate more efficiently, reduces the risk of maceration and blistering. It is also comfortable and eliminates pain on removal. Besides, SIS is also made available in the form of gel for intra site use. Dressings made up of a 3D crosslinked starch matrix with 0.9% iodine as IODOSORB and non-woven cellulose mixed with gelling cellulose ethyl sulphonate (CES) fibers as DURAFIBER dressings are also available from the same company.

4.5.15 Cyto-Matrix

Cyto-matrix, though it sounds like a 3D scaffold for cells, is not a scaffold but a natural health product company from Canada (http://www.cyto-matrix.com). Most of its products are formulated using different combinations of natural products. Inflammatrix®, one of the products for example, is a comprehensive blend of advanced enzyme and herbals with added benefits of Broelain and Quecetin. This particular formulation is stated to exhibit anti-inflammatory, fibrinolytic, mucolytic properties besides inhibiting platelet aggregation. It is said to support respiratory and arterial health and helps in wound healing. IM-Matrix™ is another formu-lation of botanical extracts that is said to have potent anti-viral, anti-bacterial and anticatarrhal properties. Basically matrix in Cyto-Matrix refers to a botanical mixture or blend.

4.5.16 AmnioGraft®

AmnioGraft® marketed by Biotissue (www.biotissue.com) is the amniotic membrane tissue which is used by eye surgeons to protect, repair and heal damaged eye surfaces. The amniotic membrane is part of the placenta that connects the growing fetus with the mother. It is the tissue that supplies nutritional ingredients required for the development of the baby in the womb. The products offered by Biotissue are cryo-preserved human amniotic membranes. Designated as Human cell, Tissue, and Cellular and Tissue-based products (HCT/P) by the U.S. Food and Drugs Administration (FDA), these products are minimally manipulated and produced in accordance with FDA regulations for Good Tissue Practices (21 CFR 1270, 1271). Placenta is collected from healthy mothers following elective Cesarean Section delivery after informed consent and extensive screening based on social habits and medical history. The donors are also screened for viruses like HIV, Hepatitis B and C, Syphilis, and West Nile using one or more antibodies.

A biologic ocular transplantation graft AmnioGraft® is used by eye doctors to treat ocular surface indications such as keratitis, corneal ulcers, SPK, pterygium, conjunctivochalasis (CCh) dry eye and Stevens-Johnson's Syndrome. It serves as a tissue replacement by delivering the unique healing actions of a cryopreserved amniotic membrane.

Bio-Tissue® also develops AmnioGuard®, a biologic glaucoma shunt graft that is used by glaucoma specialists to cover a wide variety of glaucoma drainage devices. PROKERA® is another device created using processed amniotic membrane for corneal bandage developed by Biotissue®. This represents a group of devices developed to heal and treat eye diseases such as keratitis, common dry eye, recurrent corneal erosions, filamentary keratitis, persistent epithelial defects, neurotrophic corneas, herpetic ulcers and other ocular surface diseases. It is approved by FDA and said to reduce inflammation while promoting regenerative healing of the ocular surface.

4.5.17 Artiss/Tisseel

Artiss or Tisseel represents a technology based on the fibrin matrix proposed to be used for hemostasis in patients undergoing surgery. It controls the bleeding in a natural way wherein the fibrinogen is converted to fibrin matrix through the thrombin enzyme as it occurs in a wounded environment. This is recommended as an alternative to check the bleeding when conventional surgical techniques like suturing, ligature and cautery are ineffective or impractical. Tisseel-Fibrin sealant is a 2 component fibrin sealant where the sealer protein solution contains human fibrinogen and a synthetic fibrinolysis inhibitor, aprotinin, which prevents premature degradation of a fibrin clot and the thrombin solution that contains human thrombin and calcium chloride. When mixed together, it mimics the final stages of the body's natural cascade of blood clotting and forms a rubbery mass that adheres and seals the wound surface at the micro level [43].

Tisseel is effective in heparinized patients. It is recommended for topical use only. With its origin from the native coagulation mechanism, it should never be injected directly into the circulatory system or into a highly vascularized tissue. These serious concerns bring limitations such as its use only by experts and only under situations where other alternatives (like stitches, bands or heat) cannot be used. By virtue of its mechanism of rendering hemostasis, the appropriateness of the dose should also be carefully considered; otherwise it can result in life threatening thromboembolic events. It should not be used in individuals with a known hypersensitivity to aprotinin or otherwise allergic and sensitive. It cannot be used for the treatment of severe or brisk arterial or venous bleeding (www.tisseel.com).

TISSEEL can be used only by experts after adequate assessment of the patient's condition, medical history and overall sensitivity. Its use is thus recommended only as topical or as an adjunct to the standard surgical techniques for reinforcement.

TISSEEL, the fibrin sealant, has been found particularly useful for hemostasis during anastomoses and cardiac surgeries. TISSEEL VH showed successful hemostasis in 5 minutes which is significantly shorter than in control topical agents used in patients undergoing restemotomy or reoperation after cardiac operations [42]. The fibrin sealant is shown to be safe and highly effective in controlling localized bleeding in cardiac reoperations. It was found effective in fully heparinized patients. A randomized single center trial showed that TISSEEL used on suture lines of colonic anastomoses could prevent incidences of complications due to leakage. TISSEEL is also found useful in managing splenic trauma. A single centre trial in comparison to historical control in patients undergoing laparotomy for blunt or penetrating traumatic injury to the spleen and/or liver showed 100% salvation of spleens in the group where TISEEL is applied compared to 36% in the historic control group.

Though TISSEEL technology works through the native mechanism yet due to the finite risks associated with it, the formulation is useful only in the hands of expert clinicians who can assess its zone of limitations (http://www.tisseel.com/us/detailed important risk information.html) as a product and also their patients' clinical need equally well. The Fibrin Sealant (Human) is being sold in frozen and lyophilized formulations under the name ARTISS (4 IU/ml) and TISSEEL (500 IU/ml) by Baxter Healthcare Corporation. They differ only in their final thrombin concentrations. The fibrin sealant being made from human plasma, though tested and treated to reduce the risks of it containing viruses and other infectious agents, it may, however in remote possibility may, still transmit them or cause allergic reactions in some patients. Another concern related to this technology is the observation that blood clotting and subsequent fibrin formation are factors for tumor cell seeding in the lungs [44]. A recent study demonstrated the implication of a fibronectin matrix in kidney tumor metastasis as well [45].

4.5.18 ECM Analog®

ECM Analog® represents a platform technology that addresses the biological and engineering needs of the biomaterial scaffold for 3D culture. It provides a proteinaceous flexible environment analogous to the extracellular matrix. The technology is amenable to incorporate the desired porosity, flexibility and wide range of other functionalities within the scaffold. The ECM Analog® matrix is unique in terms of simplicity of handling [46]. It can be stored at room temperature for 5 years or more without any change in its constitution. Although created from a natural polymer, it can withstand sterilization through autoclave, the most conventional method in a cell culture lab. Furthermore, the optimum porosity and transparency of the scaffold hydrogel makes it amenable to undertake microscopic examinations, using the ordinary inverted microscope. It is easy to differentiate methylene blue stained cells from the rest of the scaffold as the ECM Analog® scaffold remains unstained (Fig. 4.6).

This cell interactive scaffold is available for research purposes in different formats through ExCel Matrix Biological Devices P Ltd (www.excellmatrix.com). The Porous Microcarrier PMc® particles of different sizes are useful for suspension culture (Fig. 4.7). ECM Graft® is meant for in vivo implant study. Discs and inserts for grafting are devised to study cellular behavior/response under an

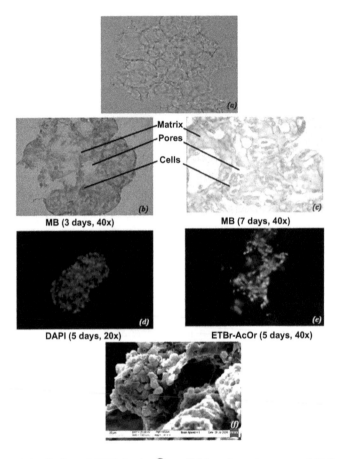

Figure 4.6 Hydrated ECM Analog® scaffold under microscope (40x) (a); monkey kidney cells (vero cell line) stained with methylene blue (MB) after 3 days' culture (b); after 7 days' culture (c); stained with DAPI (d) and ethidium bromide and acridine orange (ETBr-AcOr) after 5 day's culture (e) and with cultured bone-marrow cells (f).

in vivo setting. Cells can be inoculated on the disc and inserted in a desired location; it can be retrieved later to evaluate the changes incurred when exposed to the in vivo environment. This is one of the unique tools which can also allow to study the fate of secretory cells (with the capability of releasing metabolically important molecules). The company also supplies non-adherent culture dishes for spheroid culture and Dot-Cult® and Confo-Cult®

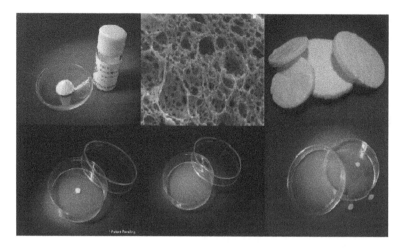

Figure 4.7 Different product formats of the ECM Analog® scaffold.

for easy confocal microscopy after culturing cells in 3D beads placed on a culture plate or a cover glass slide.

The ECM Analog® scaffold with controllable porosity and cell interactivity has a potential for in vivo use. The efficacy and safety of ECM Graft® has already been proven in animals, suggesting its viability in organotypic culture and tissue engineering. Being micro- and nanoporous, the scaffold could also be made viable as a drug carrier. Its compatibility with cells makes it useful even for carrying normal and/or recombinant cell types.

4.6 Future Trends/Challenges Ahead

Insights into healthy and diseased organ functions are acquired through qualitative evaluations of cellular responses in structurally intact and complex organ biopsies. Knowledge gained by cell biologists, pathologists and clinicians over the years has led us to believe that the ultimate decision of a cell to differentiate, proliferate, migrate, undergo apoptosis or perform other specific functions is a well-coordinated response to its molecular interaction with the surrounding ECM components and neighbouring cells. It is observed that ECM regulates cellular morphology and biochemical signals which in turn control the transcription of genes associated with specialized functions. Tissue-specific ECM through coordinated interactions of its multivalent constituents provides an array of ligands for integrins, the cell-surface receptors. These receptor-ligand interactions supported by distinct, unique morphologies lead to specific cellular functions. Understanding ECM-cell interaction is, therefore, imperative for many possible therapeutic interventions during critical tissue damage caused by trauma or as a pathological consequence.

An ideal ex vivo culture scaffold should recapitulate both the architecture and differentiated functions of a given organ/tissue while allowing experimental intervention. By doing so, cell-based models may facilitate systematic and quantitative analyses that address at the molecular level questions like how normal organ structure and functions are maintained or how the balance is lost during gradual progression towards diseases [47]. Thus, creating a biodegradable 3D scaffold with existing possibilities of incorporating biochemical cues for orchestration of a functional tissue is the fundamental need in developing an efficient ECM substitute. An ECM mimicking 3D cell culture technology can, however, be exploited to its full potential only if systematic efforts are invested in advancing the missing links like: thorough understanding of cellular response to the nature and density of ECM constituents, of mechanisms involved in cell differentiation; and characterization and cataloguing of molecular agents that can distinctly influence cell–cell interaction, induce ECM modulation or stimulate cell migration, proliferation, differentiation and apoptosis in an ex vivo environment.

Due to the unprecedented nature of research and development on the 3D scaffold, sophisticated tools and technologies are also required at different levels of development, analysis, standardization and validation of such materials. New advanced evaluation techniques or substantial modifications in older ones may be requisite for quantification and confirmation of biomarkers that could define various growth and differentiation stages in an ex vivo 3D environment. In other words, multi-parametric approaches are essential in defining the physiological status of growing cells with credible accuracy.

Current technologies allow us to engineer only those tissues which are simple in terms of their structural and functional complexities. Thus, several challenges remain to be addressed before TE and RM become acceptable as the new modes of treatment. These include challenges in creating the apt architecture associated with native tissues and organs; mimicking the complex chemical and biochemical functionalities with respect to natural ECM; and understanding cell-ECM dynamics and its impact on ECM modulation. Thus, appropriate molecular orchestration is the key in creating efficient polymeric scaffolds. Characterization and establishing the status (age and state) of seed cells are also some of the important issues in ex vivo TE, especially when primary or stem cells are intended to be used. The success of a scaffold-based RM will greatly depend on its ability to sensitize the in vivo microenvironment in which it is placed. The scaffold should have the ability to present an appropriate biochemical template with essential microstructural complexity. Only then it can attract and facilitate the healthy cells to be resident for assuming the lost or degenerated physiological function. Healthy integration with the rest of the system is also an essential requirement even for bone and cartilage tissue implants. Some of the challenges associated with TE and RM and applicable biomechanics of the scaffold are discussed by Dutta and Dutta [48] and also by Butler et al. [23]. For example, in some cases the implant should be able to re-establish vascularization and in others it may just have to integrate with the pre-existing ECM. Tourovskaia et al. have tried to address the issue of angiogenesis by creating a scaffold from a decellularized porcine small bowl segment with preserved tubular structures of the capillary network [49]. Human vasculature has also been modeled by generating multilayered tubes of smooth muscle cells and subsequent luminal seeding of endothelial cells [50].

A number of companies have indulged in developing implantable tissue engineered products. These products are aimed to replace or restore human tissue functions by combining engineering principles and materials. They may be devised to carry and dispense the living cells in a viable manner. Many of the products and devices that have been developed and are in a commercially advanced stage recommend the use of autologous cells for obvious reasons. They are available as cell free scaffolds and may even be used as fillers to acquire the desired shape or as a graft of a defined shape. They may also be inoculated with host cells to foster the regeneration of a patient's own tissue.

As of now the whole tissue transplants that could be successfully engineered ex vivo include blood vessels, full thickness skin and bladders. Thus there exist three major types of implants:

(1) Cell free scaffolds or support matrix that could be injected or implanted. They generally are constructed using tissue ECM ingredients for better compatibility. Some of the companies developing products with this approach are 3DM, Cytomatrix, Celltrix, Forticell Bioscience, Baxter, Cook Biotek, Fidia, Integra, Lifecell, ExCell Matrix, Medtronix, Orthovita and so on.

(2) Cell encapsulated scaffolds or matrix where the substrate acts more like a cell carrier. They may be single-cell aggregates or in sheet format, which could assist the partially damaged tissue. In other words, they can be used as an assistive device. BioEngine, GeneGrafts, Microislet, HepaLife, LCT, Neurotech, NsGene, TiGenix, Vital Therapies, Arthrokinetics, Biotissue technology, Cell Matrix, Advanced Biohealing, CellTran, Cell Matrix, Genzyme, Hybrid Organ, Interface Biotech, Organogenesis, Vasotissue Technologies are some of the companies using this approach.

(3) Whole tissue implants which generally are organized ex vivo into an appropriate shape through a cell-substrate combination. Artificially created blood vessels, cartilage, bone and bladders are some of the examples of such tissue implants. Bio Nova, Humacyte, Cytograft, Educell, Histogenics, Intercytex, ISTO, Tengion, Theregen, Euroderm, MatTek, Karocell Tissue Engineering, Skin Ethic Laboratories are some of the companies mentioned which follow the whole tissue approach [51].

Disclaimer: As the authors Ranjna C Dutta and Aroop K Dutta are not responsible for any information that may be considered as misrepresentation or deviation from actual products or technology of any company that finds contextual mention in the Chapter 4 of this book. The details provided here are based solely on the information available on respective company's website and/or literature at the time of writing this manuscript.

References

1. Caicedo-Carvajal CE, Liu Q, Remache Y, Goy A and Suh KS (2011). Cancer tissue engineering: a novel 3D polystyrene scaffold for in vitro isolation and amplification of lymphoma cancer cells from heterogeneous cell mixtures. *J Tissue Eng*, **2011**, 362326.

2. Altmann B, Welle A, Giselbrecht S, Truckenmüller R and Gottwald E (2009). The famous versus the inconvenient - or the dawn and the rise of 3Dculture systems. *World J Stem Cells*, **1**, 43–48.

3. Dutta RC and Dutta AK (2012). ECM analog technology: a simple tool for exploring cell-ECM dynamics. *Front Biosci (Elite Ed)*, **4**, 1043–1048.

4. Dutta RC and Dutta AK (2010). Comprehension of ECM-cell dynamics: a prerequisite for tissue regeneration. *Biotechnol Adv*, **28**, 764–769.

5. Dutta RC and Dutta AK (2009). Cell interactive 3D scaffold; advances and applications. *Biotechnol Adv.* **27**, 334–39, (2009)

6. Dutta RC (2011). In search of optimal scaffold for regenerative medicine and therapeutic delivery. *Ther Deliv*, **2**(2), 231–234.

7. Read JA, Winter VJ, Eszes CM, Sessions RB and Brady RL (2001). Structural basis for altered activity of M and H-isozyme forms of human lactate dehydrogenase. *Proteins*, **43**, 175–185.

8. Cukierman E, Pankov R, Stevens DR and Yamada K (2001). Taking cellmatrix adhesions to the third dimension. *Science*, **294**, 17081712.

9. Engler AJ, Sen S, Sweeney HL and Discher DE (2006). Matrix elasticity directs stem cell lineage specification. *Cell*, **126**, 677–689.

10. Keung AJ, Healy KE, Kumar S and Schaffer DV (2010). Biophysics and dynamics of natural and engineered stem cell microenvironments. *Wiley Interdiscip Rev Syst Biol Med*, **2**, 49–64.

11. Ji C, Khademhosseini A and Dehghani F (2011). Enhancing cell penetration and proliferation in chitosan hydrogels for tissue engineering applications. *Biomaterials*, **32**, 9719–9729.

12. Nichol JW and Khademhosseini A (2009). Modular tissue engineering: engineering biological tissues from the bottom up. *Soft Matter*, **5**, 1312-1319.

13. Nuernberger S, Cyran N, Albrecht C, Redl H, Vecsei V, et al. (2011). The influence of scaffold architecture on chondrocyte distribution and behavior in matrix-associated chondrocyte transplantation grais. *Biomaterials*, **32**, 1032–1040.

14. Martina M and Hutmacher DW (2007). Biodegradable polymers applied in tissue engineering research: a review. *Polym Int*, **57**, 145–157.

15. Aurand ER, Wagner J, Lanning C and Bjugstad KB (2012). Building biocompatible hydrogels for tissue engineering of the brain and spinal cord. *J Funct Biomater*, **3**, 839–863.

16. Garg T, Singh O, Arora S and Murthy R (2012). scaffold: a novel carrier for cell and drug delivery. *Crit Rev Her Drug Carrier Syst*, **29**, 1–63.

17. Malafaya PB, Silva GA and Reis RL (2007). Natural-origin polymers as carriers and scaffolds for biomolecules and cell delivery in tissue engineering applications. *Adv Drug Deliv Rev*, **59**, 207–233.

18. Peter SJ, Miller MJ, Yasko AW, Yaszemski MJ and Mikos AJ (1998). Polymer concepts in tissue engineering. *J Biomed Mater Res*, **43**, 422–427.

19. Marra KG, Szem JW, Kumta PN, DiMilla PA and Weiss LE (1999). In vitro analysis of biodegradable polymer blend/hydroxyapatite composites for bone tissue engineering. *J Biomed Mater Res*, **47**, 324–335.

20. Lenhert S, Meier MB, Meyer U, Chi L and Wiesmann HP (2004). Osteoblast alignment, elongation and migration on grooved polystyrene patterned by Langmuir-Blodgett lithography. *Biomaterials*, **26**, 563–570.

21. Carrel A and Lindberg C (1938). The culture of organs. *Can Med Assoc J*, New York: Paul B. Hoeber, Inc., Medical Book Department of Harper & Brothers.

22. Skalak R and Fox C (1988). *NSF Workshop: UCLA Symposia on Molecular and Cellular Biology*. Molecular and Cellular Biology, New Series. New York: Alan R. Liss, Inc.

23. Butler DL, Goldstein SA, Guldberg RE, Guo XE, Kamm R, Laurencin CT, McIntire LV, Mow VC, Nerem RM, Sah RL, Soslowsky LJ, Spilker RL and Tranquillo RT (2009). The impact of biomechanics in tissue engineering and regenerative medicine. *Tissue Eng Part B Rev*, **15**(4), 477–484.

24. Spiro RG (1960). Characterization and quantitative determination of the hydroxylysine-linked of carbohydrate units of several collagens. *J Biol Chem*, **244**(3), 602–612.

25. Kleinman HK and Martin GR (2005). Matrigel: basement membrane extracellular matrix with biological activity. *Semin Cancer Biol*, **15**, 378–386.

26. Hughes CS, Postovit LM and Lajoie GA (2010). Matrigel: a complex protein mixture required for optimal growth of cellculture. *Proteomics*, **10**(9), 1886–1890.

27. Benton G, George J, Kleinman HK and Arnaoutova I (2009). Advancing science and technology via 3D culture on basement membrane matrix. *J Cell Physiol*, **221**(1), 18–25.

28. Benton G, Kleinman HK, George J and Arnaoutova I (2011). Multiple uses of basement membrane-like matrix (BME/Matrigel) in vitro and in vivo with tumor cells. *Int J Cancer*, **128**(8), 1751–1757.

29. Knight E, Murray B, Carnachan R and Przyborski S (2011). Alvetex®: polystyrene scaffold technology for routine three dimensional cell culture. *Methods Mol Biol*, **695**, 323–340.

30. Godugu C, Patel AR, Desai U, Andey T, Sams A and Singh M (2013). AlgiMatrix™ based 3D cell culture system as an in-vitro tumor model for anticancer studies. *PLoS One,* **8**(1), e53708.

31. Loo Y, Zhang S and Hauser CAE (2012). From short peptides to nanofibers to macromolecular assemblies in biomedicine. *Biotechnol Adv*, **30**, 593–603.

32. Parcells AL, Karcich J, Granick MS and Marano MA (2014). The use of fetal bovine dermal scaffold (PriMatrix) in the management of full-thickness hand burns. *Eplasty*, **14**, e36.

33. Migliore A, Massafra U, Bizzi E, Vacca F, Martin-Martin S, Granata M, Alimonti A and Tormenta S (2009). Comparative, double-blind, controlled study of intra-articular hyaluronic acid (Hyalubrix®) injections versus local anesthetic in osteoarthritis of the hip. *Arthritis Res Ther*, **11**, R183.

34. Yang J, Yamato M, Nishida K, Ohki T, Kanzaki M, Sekine H, Shimizu T and Okano T (2006). Cell delivery in regenerative medicine: the cell sheet engineering approach. *J Control Release,* **116**(2), 193–203.

35. Schanz J, Puscha J, Hansmanna J and Walles H (2010). Vascularised human tissue models: a new approach for the refinement of biomedical research. *J Biotechnol*, **148**, 56–63.

36. Steinke M, Dally I, Friedel G, Walles H and Walles T (2015). Host-integration of a tissue-engineered airway patch: two-year follow-up in a single patient. *Tissue Eng Part A*, **21**(3–4), 573–579.

37. Groeber F, Engelhardt L, Lange J, Kurdyn S, Schmid FF, Rücker C, Mielke S, Walles H and Hansmann J (2016). A first vascularized skin equivalent as an alternative to animal experimentation. *ALTEX*, **33**(4), 415–422.

38. Darr A and Calabro A (2009). Synthesis and characterization of tyramine-based hyaluronan hydrogels. *J Mater Sci Mater Med*, **20**(1), 33–44.

39. Chin L, Calabro A, Walker E and Derwin KA (2012). Mechanical properties of tyramine substituted-hyaluronan enriched fascia extracellular matrix. *J Biomed Mater Res Part A*, **100A**, 786–793.

40. Fumoto H, Takaseya T, Shiose A, Saraiva RM, Arakawa Y, Park M, Rao S, Dessoffy R, Chen JF, Zhou Q, Calabro A Jr, Banbury M and Fukamachi K (2010). Mitral annular remodeling to treat functional mitral regurgitation: a pilot acute study in a canine model. *Heart Surg Forum*, **13**(4), E247–E250.

41. Brown-Etris M, Cutshall WD and Hiles MC (2002). A new biomaterial derived from small intestine submucosa and developed into a wound matrix device. *Wounds*, **14**(4), 150–166.

42. Rousou J, Gonzalez-Lavin L, Cosgrove D, et al. (1989). Randomized clinical trial of fibrin sealant in patients undergoing resternotomy or reoperation after cardiac operations. *J Thorac Cardiovasc Surg*, **97**, 194–203.

43. Gorkun OV, Veklich YI, Weisel JW and Lord ST (1997). The conversion of fibrinogen to fibrin: recombinant fibrinogen typifies plasma fibrinogen. *Blood*, **89**(12), 4407–4414.

44. Palumbo JS, Kombrinck KW, Drew AF, Grimes TS, Kiser JH, Degen JL, et al. (2000). Fibrinogen is an important determinant of the metastatic potential of circulating tumor cells. *Blood*, **96**, 3302–3309.

45. Knowles LM, Gurski LA, Maranchie JK and Pilch J (2015). Fibronectin matrix formation is a prerequisite for colonization of kidney tumor cells in fibrin. *J Cancer*, **6**(2), 98–104.

46. Dutta RC and Dutta AK (2012). ECM analog technology: a simple tool for exploring cell-ECM dynamics. *Front Biosci (Elite Ed)*, **4**, 1043–1048.

47. Polyakova V, Miyagawa S, Szalay Z, Risteli J and Kostin S (2008). Atrial extracellular matrix remodelling in patients with atrial fibrillation. *J Cell Mol Med*, **12**(1), 189–208.

48. Dutta RC and Dutta AK (2014). Strategic challenges in practicing tissue engineering and regenerative medicine. *J Regen Med*, **3**, 1.

49. Tourovskaia A, Fauver M, Kramer G, Simonson S and Neumann T (2014). Tissue-engineered microenvironment systems for modeling human vasculature. *Exp Biol Med (Maywood)*, **239**, 1264–1271.

50. Griffith LG and Swartz MA (2006). Capturing complex 3D tissue physiology in vitro. *Nat Rev Mol Cell Biol*, **7**, 211–224.

51. Khademhosseini A, Vacanti JP and Langer R (2009). Progress in tissue. *Sci Am*, **300**(5), 64–71.

Recommended Reading

(i) Silver FH (1994). *Biomaterials, Medical Devices and Tissue Engineering: An Integrated Approach.* Chapman & Hall, London.

(ii) Comper WD, ed. (1996). *Extracellular Matrix, Vol. 1: Tissue Function.* Amsterdam, the Netherlands: Harwood Academic Publishers.

(iii) Comper WD, ed. (1996). *Extracellular Matrix, Vol. 2: Molecular Components and Interactions.* Amsterdam, the Netherlands: Harwood Academic Publishers.

(iv) Cooper GM (2000). *The Cell: A Molecular Approach*, 2nd edn. Sunderland (MA): Sinauer Associates.

(v) Lanza RP, Langer RS and Vacanti JP (2000). *Principles of Tissue Engineering*, 2nd edn. San Diego: Academic Press.

(vi) Guilak F, Butler DL and Goidstein SA (2004). *Functional Tissue Engineering.* New York: Springer Verlag.

(vii) Miner JH (2005). *Extracellular Matrix in Development and Disease*, in series of Advances in Developmental Biology. Oxford: Elsevier

(viii) Ricard-Blum S (2011). The collagen family. *Cold Spring Harb Perspect Biol*, **3**, a004978.

(ix) Khang G, Kim MS and Lee HB (2007). *Manuals in Biomedical Research, Vol 4: A Manual for Biomaterials/Scaffold Fabrication Technology.* Singapore: World Scientific Publishing.

(x) RA Preti (2005). Bringing safe and effective cell therapies to the bedside. *Nat Biotechnol*, **23**(7), 801–804.

(xi) Khademhosseini A, Langer R, Borenstein J and Vacanti JP (2006). Microscale technologies for tissue engineering and biology. *Proc Natl Acad Sci USA*, **103**(8), 2480–2487.

(xii) Mikos AG, et al. (2006). Engineering complex tissues. *Tissue Eng*, **12**(12), 3307–3339.

(xiii) Lysaght MJ, et al. (2008). Great expectations: private sector activity in tissue engineering, regenerative medicine, and stem cell therapeutics. *Tissue Eng Part A*, **14**(2), 305–315.

Glossary

2D	Two-dimensional
3D	Three-dimensional
AcOr	Acridine orange (*a fluorescent dye that stains both DNA and RNA often used for cell cycle determination*)
ASTM	American Society for Testing and Materials
BM	Basement membrane
BMSC	Bone marrow stem cells
Biocompatible	No adverse effects on biological system
Biomaterial	Biocompatible material for in vivo use
Bioinert	No interaction with biological system
Bioerodible/biodegradable	Dissolves or degrades in biological system
overtime CAD	Compuer-aided design
CAM	Cell adhesion molecules
CaP	Calcium phosphate
CS	Chondroitin sulfate
DAPI	(*4',6-diamidino-2-phenylindole; is a fluorescent* stain *that binds strongly to double stranded DNA and is used extensively in fluorescence microscopy*)
DS	Dermatan sulfate
ECM	Extracellular microenvironment/matrix

EMILIN	Elastin microfibril interface located protein
ECM	Extracellular microenvironment/matrix
EGF	Epidermal/epithelial growth factor
ETBr	Ethidium bromide (*anintercalating agent commonly used as a fluorescent tag/nucleic acid stain*)
FACIT	Fibril-associated collagen with interrupted triple helix
FBF	Fibroblast growth factor
FN	Fibronectin
GAG	Glycosaminoglycan
GBM	Glomerular basement membrane
GP	Glycoprotein
HA	Hyaluronic acid
HaP/HAP	Hydroxyapatite
HS	Heparan sulfate
HSP	Heat shock protein
HSPG	Heparan sulfate proteoglycan
KS	Keratinn sulfate
LEM	Linear elastic modulus
LM	Laminin
LOX	Lysyl-oxidase
LTBP	Latent transforming growth factor-beta-binding protein
MAGP	Microfibril associated glycoproteins
MDCK	Madin-Darby canine kidney cells
MMP	Matrix metalloproteinase
NC	Non-collagenous
Pa	Pascal
PCL	Polycaprolactone
PCM	Pericellular matrix
PEDOT	Poly(3,4-ethylenedioxythiophene)
PEG	Polyethylene glycol

PG	Proteoglycan
PGA	Polyglycolic acid
PLGA	Poly lactic-co-glycolic acid
PSC	Pluripotent stem cells
PSS	Poly(4-styrene sulfonate)
PVA	Poly vinyl alcohol
QFDE	Quick freeze deep-etch
RM	Regenerative medicine
RGD	Arg-Gly-Asp (a tripeptide)
SLRP	Small leucin rich proteoglycans
SMC	Smooth muscle cells
TCP	Tricalcium phosphate
TE	Tissue engineering
TMP	Trans-membrane proteoglycans
TGF-β	Transforming growth factor beta
Tensigrity	Tensigrity represents tensional integrity or floating compression. It is the characteristic property of a stable three-dimensional structure that consists of components under tension that are contiguous, and components under compression. An architectural principle conceived by R. Buckminister Fuller, in which compression and tension are used to give a structure its form and shape that was adopted by Ingber to explain cell stability through forces of compression and tension in the cytoskelatal components.

Index